Hazardous Materials

Awareness and Operations

THIRD EDITION

Rob Schnepp

Student Workbook

Jones & Bartlett Learning
World Headquarters
5 Wall Street
Burlington, MA 01803
978-443-5000
info@jblearning.com
www.jblearning.com

National Fire Protection Association
1 Batterymarch Park
Quincy, MA 02169
www.NFPA.org

International Association of Fire Chiefs
4025 Fair Ridge Drive
Fairfax, VA 22033
www.IAFC.org

Jones & Bartlett Learning books and products are available through most bookstores and online booksellers. To contact Jones & Bartlett Learning directly, call 800-832-0034, fax 978-443-8000, or visit our website, www.jblearning.com.

Substantial discounts on bulk quantities of Jones & Bartlett Learning publications are available to corporations, professional associations, and other qualified organizations. For details and specific discount information, contact the special sales department at Jones & Bartlett Learning via the above contact information or send an email to specialsales@jblearning.com.

Production Credits
General Manager: Kimberly Brophy
VP, Product Development and Executive Editor: Christine Emerton
Executive Editor: Bill Larkin
Vendor Manager: Nora Menzi
VP, Sales, Public Safety Group: Matthew Maniscalco
Director of Marketing Operations: Brian Rooney
VP, Manufacturing and Inventory Control: Therese Connell
Composition and Project Management: S4Carlisle Publishing Services
Cover Design: Kristin E. Parker
Text Design: Kristin E. Parker
Rights & Media Specialist: Thais Miller
Media Development Editor: Shannon Sheehan
Cover Image: © Jones & Bartlett Learning. Photographed by Glen E. Ellman.
Printing and Binding: Edwards Brothers Malloy
Cover Printing: Edwards Brothers Malloy

Editorial Credits
Authors: Rudy Horist and Matt Muldoon

ISBN: 978-1-284-14694-3

6048

Printed in the United States of America
22 21 20 19 18 10 9 8 7 6 5 4 3 2 1

Contents

Note to the student: Consult your instructor for access to the Student Workbook Answer Key.

Regulations, Standards, and Laws

Workbook Activities

The following activities have been designed to help you. Your instructor may require you to complete some or all of these activities as a regular part of your hazardous materials training program. You are encouraged to complete any activity your instructor does not assign you as a way to enhance your learning in the classroom.

Chapter Review

The following exercises provide an opportunity to refresh your knowledge of this chapter.

Matching

Match each of the terms in the left column to the appropriate definition in the right column.

_____	**1.** Awareness level	**A.**	Training that provides the ability to control a release using specialized clothing and control equipment
_____	**2.** Technician level	**B.**	The impure substance that remains after a process or manufacturing activity
_____	**3.** Hazardous material	**C.**	OSHA document that contains the hazardous materials response competencies
_____	**4.** HAZWOPER	**D.**	An occupancy or facility that presents a high potential for loss of life
_____	**5.** Hazardous waste	**E.**	The organization that develops and issues voluntary consensus-based standards
_____	**6.** NFPA	**F.**	Training that provides the ability to recognize a potential hazardous emergency and secure the area
_____	**7.** Target hazard	**G.**	Any material that is capable of posing an unreasonable risk to human health, safety, or the environment
_____	**8.** SARA	**H.**	A committee that disseminates information about hazardous materials to the public
_____	**9.** LEPC	**I.**	The lowest level of training that qualifies the individual to respond to a hazardous materials incident
_____	**10.** Operations level	**J.**	Legislation that created a standard practice for a community so it can understand and be aware of the chemical hazards in the community

CHAPTER

1

Multiple Choice

Read each item carefully, and then select the best response.

_____ **1.** A material that poses an unreasonable risk to the health and safety of the public and/or the environment if it is not controlled properly during handling, processing, and disposal is called a:
A. hazardous waste
B. hazardous material
C. hazardous target
D. hazardous substance

_____ **2.** Which of the following organizations develops and issues voluntary consensus standards?
A. National Fire Protection Agency (NFPA)
B. Occupational Safety and Health Administration (OSHA)
C. United States Department of Transportation (DOT)
D. Environmental Protection Agency (EPA)

_____ **3.** In the United States, the federal document containing the hazardous materials response competencies is known as the:
A. Environmental Protection Response Act (EPRA)
B. Hazardous Waste Operations and Emergency Response (HAZWOPER)
C. Code of Federal Regulations (CFR)
D. National Fire Protection Association (NFPA) Standard 472

_____ **4.** The individual responsible for all incident activities, including the development of strategy and tactics, is training to the ________ level.
A. operations
B. technician
C. specialist
D. Incident Command

_____ **5.** What act requires a business that handles chemicals to report the storage methods, storage type, and quantity to the fire department and the local emergency planning committee?
A. Superfund Amendments and the Reauthorization Act
B. Local Emergency Planning Committee Act
C. Emergency Planning and Community Right to Know Act
D. Occupational Safety and Health Act

_____ **6.** Which of the following is a group that gathers information about hazardous materials and disseminates that information to the public?
A. LEPC
B. NFPA
C. SERC
D. MSDS

______ 7. Which of the following is a state organization that acts as a liaison between local and state-level response authorities?
- A. SARA
- B. EPA
- C. SERC
- D. NFPA

______ 8. Which competency level is trained to recognize a hazardous materials emergency and call for assistance?
- A. awareness
- B. operations
- C. technician
- D. specialist

______ 9. What response level is trained to take defensive actions at a hazardous materials incident?
- A. awareness
- B. operations
- C. technician
- D. specialist

______ 10. Which is the minimum level of training to take offensive actions at a hazardous materials incident?
- A. awareness
- B. operations
- C. technician
- D. specialist

Vocabulary

Define the following terms using the space provided.

1. Local Emergency Planning Committee (LEPC)

____

____

____

2. Hazardous waste

____

____

____

3. Target hazard

____

____

____

4. HAZWOPER

____

____

____

5. Operations level

____

____

____

Fill-In

Read each item carefully, and then complete the statement by filling in the missing word(s).

1. The model codes and standards of the United Nations use the term ________________ to describe hazardous materials.

2. The acronym CBRNE stands for ________________, ________________, ________________, ________________, and ________________.

3. Documents that are issued and enforced by governmental bodies are ________________.

4. ________________ activities enable agencies to develop logical and appropriate response procedures for anticipated incidents.

5. Individuals trained to the ________________ level are considered personnel and not responders.

6. Hazardous materials ________________ are trained to use offensive actions.

7. NFPA ________________ is the *Standard for Competence of Responders to Hazardous Materials/Weapons of Mass Destruction Incidents.*

8. The federal document containing the hazardous materials response competencies is known as ________________.

9. The material that remains after a manufacturing plant has used some chemicals and they are no longer pure is known as ________________.

10. States that have adopted OSHA safety and health regulations are called ________________ states.

True/False

If you believe the statement to be more true than false, write the letter "T" in the space provided. If you believe the statement to be more false than true, write the letter "F."

1. ______ Hazardous materials incidents are handled in a more deliberate fashion than structural firefighting.

2. ______ The NFPA issues regulations that govern hazardous materials response.

3. ______ Each state has a State Emergency Response Commission (SERC) that acts as a liaison between local and state levels of authority.

4. _______ The actions taken at hazardous materials incidents are largely dictated by the chemicals involved.

5. _______ The lowest level of training that qualifies an individual to respond to hazardous materials incidents is the awareness level.

6. _______ When approaching a hazardous materials event, you should make a conscious effort to change your response perspective.

7. _______ Preplanning activities are primarily a responsibility of the Local Emergency Planning Committee (LEPC).

8. _______ Standards are documents issued and enforced by governmental agencies.

9. _______ OSHA requires that responders receive annual refresher training.

10. _______ Weapons of mass destruction include any weapons involving toxic or poisonous chemicals.

Short Answer

Complete this section with short written answers using the space provided.

1. Identify the four levels of hazardous materials training and competencies, according to NFPA 1072.

____

____

____

____

2. Discuss the role of the NFPA in developing standards for hazardous materials response.

____

____

____

____

Hazardous Material Alarms

The following real-case scenarios will give you an opportunity to explore the concerns associated with hazardous materials. Read each scenario, and then answer each question in detail.

1. You recently completed your hazardous materials awareness-level training. While driving through another community, you come across a commercial vehicle that is pulled over on the side of the road and appears to be leaking a bright green fluid from the cargo area. What would be appropriate actions for you to take?

____

____

____

____

2. You are asked to speak to a class of new-recruit firefighters regarding how the fire department approaches hazardous materials incidents. What will your discussion include?

____

____

____

____

Recognizing and Identifying the Hazards

Workbook Activities

The following activities have been designed to help you. Your instructor may require you to complete some or all of these activities as a regular part of your hazardous materials training program. You are encouraged to complete any activity that your instructor does not assign you as a way to enhance your learning in the classroom.

Chapter Review

The following exercises provide an opportunity to refresh your knowledge of this chapter.

Matching

Match each of the terms in the left column to the appropriate definition in the right column.

	Term		Definition
______	**1.** Container	**A.**	The opening in the lid of a closed-head drum
______	**2.** Dewar container	**B.**	Marking system characterized by a set of diamonds
______	**3.** Bung	**C.**	Type of shipping paper for railroad transportation
______	**4.** Carboy	**D.**	Type of shipping paper for road and highway transportation
______	**5.** NFPA 704	**E.**	Any vessel or receptacle that holds a material
______	**6.** Bill of lading	**F.**	A thermos-like container designed to maintain the appropriate temperature
______	**7.** Waybill	**G.**	The process by which chemical substances travel through body tissues until they enter the bloodstream
______	**8.** Inhalation	**H.**	A type of vessel used to transport and store corrosives
______	**9.** Absorption	**I.**	The process by which chemical substances enter the body through the respiratory system
______	**10.** Injection	**J.**	The process by which chemical substances are brought into the body through open cuts and abrasions

Multiple Choice

Read each item carefully, and then select the best response.

______ **1.** The largest database on chemical information is produced by the:

- **A.** National Fire Protection Association (NFPA)
- **B.** Chemical Abstracts Service (CAS)
- **C.** Chemical Transportation Emergency Center (CHEMTREC)
- **D.** U.S. Environmental Protection Agency (EPA)

______ **2.** Once size-up is complete, responders can begin to formulate a plan, which begins with taking basic actions. This is known as:

- **A.** Safety, Isolate, Notify (SIN)
- **B.** Calculate, Secure, Notify (CSN)
- **C.** Notify, Secure, Plan (NSP)
- **D.** Safety, Security, Resources (SSR)

CHAPTER 2

______ **3.** Drums used to store corrosives are typically made of:
- **A.** steel
- **B.** stainless steel
- **C.** fiberglass
- **D.** polyethylene

______ **4.** Drum bungs can be removed with a:
- **A.** bung wrench
- **B.** drum ratchet
- **C.** cinching wrench
- **D.** drum ring

______ **5.** A glass, plastic, or steel container that holds 5 to 15 gallons of product is a:
- **A.** bottle
- **B.** cylinder
- **C.** carboy
- **D.** drum

______ **6.** Propane cylinders contain a liquefied gas and have low pressures of approximately:
- **A.** 50–110 psi
- **B.** 150–200 psi
- **C.** 200–300 psi
- **D.** 300–400 psi

______ **7.** Gaseous substances that have been chilled until they liquefy are classified as:
- **A.** cryogenic gases
- **B.** crystals
- **C.** Dewar liquids
- **D.** cryogenic liquids

______ **8.** The NFPA 704 marking system is designed for what type of use?
- **A.** Road transportation
- **B.** Fixed facilities
- **C.** Intermodal containers
- **D.** Chemical transport vehicles

______ **9.** Within the NFPA hazard identification system, which number is used to identify materials that can cause death after a short exposure?
- **A.** 4
- **B.** 2
- **C.** 1
- **D.** 0

______ **10.** Within the NFPA hazard identification system, which number is used to identify materials that will not burn?
- **A.** 4
- **B.** 2
- **C.** 1
- **D.** 0

_____ **11.** Shipping papers for road and highway transportation are kept:
A. at the rear of the trailer
B. in each individual container within the trailer
C. in the cab of the vehicle
D. There is no specific location for them to be kept.

_____ **12.** The U.S. military marking system for hazardous materials uses an orange octagon with the number one (1) printed in the center to indicate what type of hazard?
A. mass detonation
B. explosion with fragments
C. mass fire
D. moderate fire

_____ **13.** Which DOT packaging group designation is used to represent the highest level of danger?
A. Packaging group I
B. Packaging group II
C. Packaging group III
D. Packaging group IV

_____ **14.** Of all of the various methods used to transport hazardous materials, which type is most rarely involved in emergencies?
A. rail
B. ship
C. road
D. pipeline

_____ **15.** Information about a pipeline's contents and owner is often found at the:
A. main office
B. substation
C. vent pipes
D. major intersections

_____ **16.** What is an example of a hazardous material that must be placarded regardless of the amount?
A. flammable gases
B. water reactive solids
C. flammable solids
D. combustible liquids

_____ **17.** An example of what the Emergency Response Guidebook (ERG) considers a small spill is:
A. a small leak from a 1-ton cylinder
B. a small leak from up to a 55-gallon drum
C. a spill from a number of small packages
D. a small spill from a railcar

_____ **18.** Which DOT class is used to identify flammable liquids?
A. Class 1
B. Class 2
C. Class 3
D. Class 7

_____ **19.** When chemical substances enter the human body through cuts or other breaches in the skin, this is known as:
A. absorption
B. inhalation
C. ingestion
D. injection

_____ **20.** The single most important piece of personal protective equipment that firefighters have at their disposal is:
A. self-contained breathing apparatus
B. gloves
C. helmet
D. air-purifying respirator

Labeling

Label the following diagrams with the correct terms.

1. Identify the colors and category for each diamond.

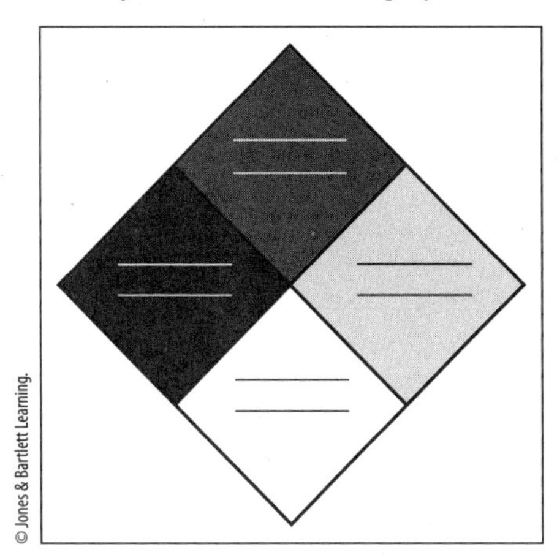

2. Identify the four ways a chemical substance can enter the body.

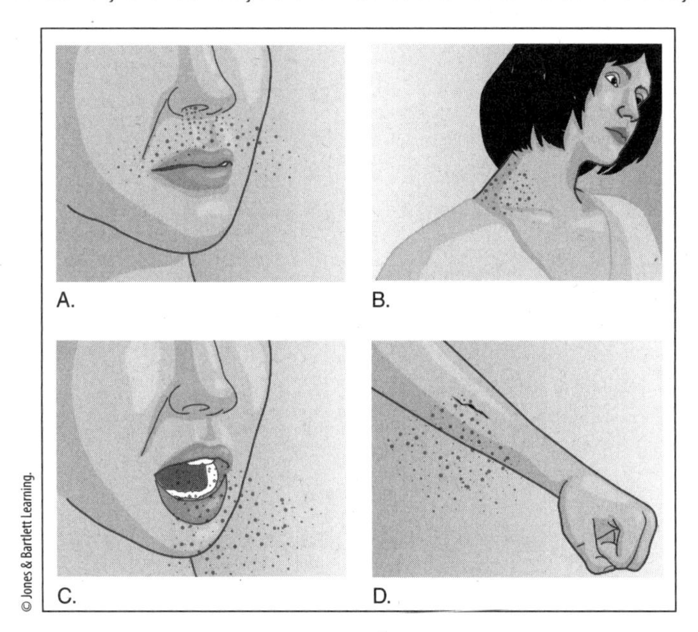

3. Identify the U.S. military system for hazardous materials.

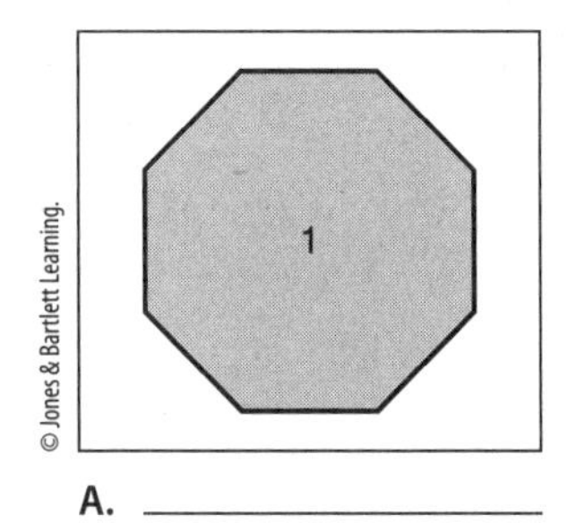

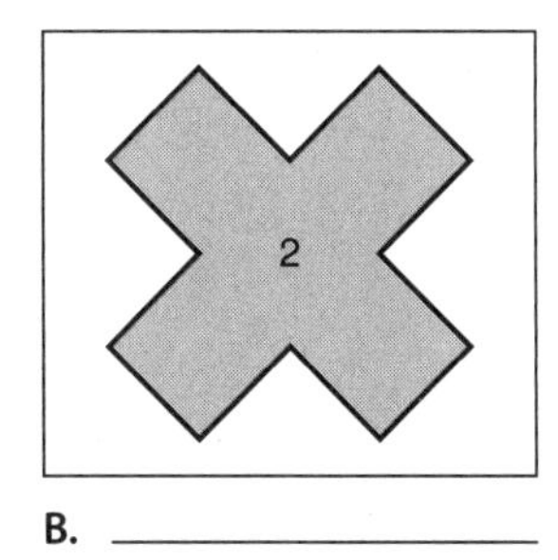

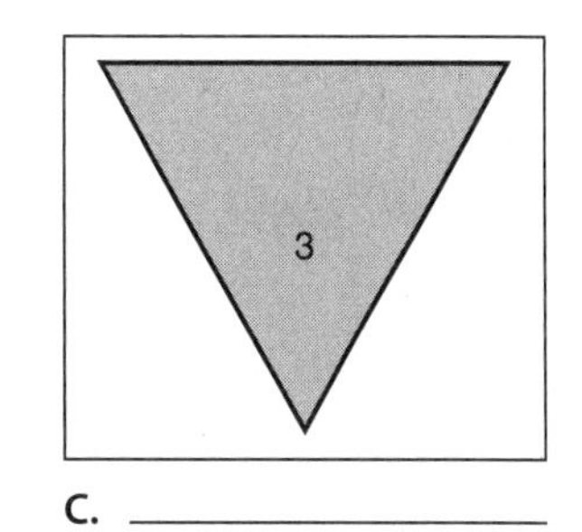

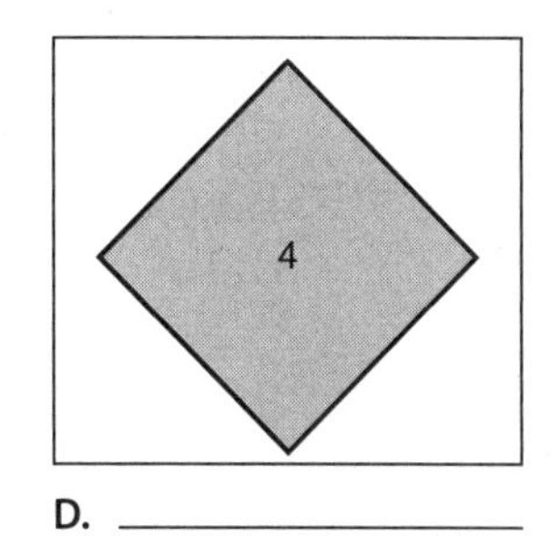

A. ____________

B. ____________

C. ____________

D. ____________

Vocabulary

Define the following terms using the space provided.

1. Shipping papers

____

____

____

2. BLEVE

____

____

____

3. Emergency Response Guidebook (ERG)

____

____

____

4. Pipeline right of way

____

____

____

5. Placards and labels

____

____

____

Fill-In

Read each item carefully, and then complete the statement by filling in the missing word(s).

1. Scene ____________________ is especially important in any hazardous materials incident.

2. Leaving potential victims inside buildings or vehicles is known as ____________________, ____________________, ____________________.

3. All facilities that use or store chemicals are required by law to have a(n) ____________________, ____________________, ____________________ on file for each chemical used or stored in the facility.

4. Large-diameter ____________________ transport natural gas, diesel fuel, and other products from delivery terminals to distribution facilities.

5. Within the NFPA hazard identification system, special hazards that react with water are identified by ____________________.

6. The ____________________ ____________________ ____________________ ____________________ describes the chemical hazards posed by a particular substance and provides guidance about personal protective equipment that employees need to use to protect themselves from workplace hazards.

7. The ____________________ marking system has been developed primarily to identify detonation, fire, and special hazards.

8. Shipping papers for railroad transportation are known as ____________________; the list of the contents of every car on the train is known as a(n) ____________________.

9. The ____________________ colored section of the Emergency Response Guidebook offers recommendations on the size and shape of protective action zones.

10. The Emergency Response Guidebook is used in the United States, ____________________, and ____________________.

True/False

If you believe the statement to be more true than false, write the letter "T" in the space provided. If you believe the statement to be more false than true, write the letter "F."

1. ______ Hazardous materials incidents can occur almost anywhere.
2. ______ Any vessel or receptacle that holds a material is known as an intermodal.
3. ______ The opening in the permanently attached lid of a closed drum is known as a bung.
4. ______ Polyethylene drums are used for corrosives and other materials that cannot be stored in steel drums.
5. ______ The NFPA 704 marking system is intended for the employees of a facility, not responders.
6. ______ OX is used to represent compressed oxygen in the NFPA hazard identification system.
7. ______ Chemical hazards in the military marking system are depicted by colors.
8. ______ The air bill is kept in the cockpit and is the pilot's responsibility.
9. ______ Placards and labels are intended to give specific information regarding the hazard inside a container or cargo tank.
10. ______ The four-digit United Nations number on a placard indicates the hazardous material's country of origin.

Short Answer

Complete this section with short written answers using the space provided.

1. List five (5) specific items of information that would typically be included on a Safety Data Sheet.

___

___

___

___

2. Identify the nine (9) Emergency Response Guidebook chemical families.

___

___

___

___

3. Identify and describe the four (4) colored sections of the Emergency Response Guidebook.

___

___

___

___

4. Describe the parts and purpose of the NFPA 704 hazard identification system.

___

___

___

___

5. Identify and define the four (4) ways through which a chemical substance can enter the human body.

___

___

___

___

Hazardous Material Alarms

The following real-case scenarios will give you an opportunity to explore the concerns associated with hazardous materials. Read each scenario, and then answer each question in detail.

1. Your engine company is first on the scene of a vehicle accident on the four-lane highway leading into town. You see a tanker truck lying on its side, with green liquid leaking from the valving. What are your initial actions regarding protecting the safety of yourself and other responders?

2. As the hazardous materials team arrives on the scene of the scenario described in Question 1, your engine company is now assigned to assist with isolating and denying entry to the scene. What would be some of the expected actions to accomplish this?

Properties and Effects

Workbook Activities

The following activities have been designed to help you. Your instructor may require you to complete some or all of these activities as a regular part of your hazardous materials training program. You are encouraged to complete any activity that your instructor does not assign you as a way to enhance your learning in the classroom.

Chapter Review

The following exercises provide an opportunity to refresh your knowledge of this chapter.

Matching

Match each of the terms in the left column to the appropriate definition in the right column.

______	**1.** Vapor	**A.**	The ability of a material to cause damage on contact to skin, eyes, or other parts of the body
______	**2.** Flash point	**B.**	An expression of a fuel/air mixture, defined by upper and lower limits, that reflects an amount of flammable vapor mixed with a given volume of air
______	**3.** Corrosivity	**C.**	Fluid build-up in the lungs
______	**4.** Toxicology	**D.**	The weight of an airborne concentration of a vapor or gas compared to an equal volume of dry air
______	**5.** Flammable range	**E.**	The gas phase of a substance
______	**6.** Vapor density	**F.**	An expression of the minimum temperature at which a liquid or solid gives off sufficient vapors such that, when an ignition source is present, the vapors will result in a flash fire
______	**7.** Pulmonary edema	**G.**	A cancer-causing agent
______	**8.** Boiling point	**H.**	The measure of the degree to which something is toxic or poisonous
______	**9.** Carcinogen	**I.**	The temperature at which a liquid will continually give off vapors in sustained amounts and, if held at that temperature long enough, will turn completely into a gas
______	**10.** Expansion ratio	**J.**	A description of a volume increase that occurs when a compressed liquefied gas changes to a gas

Multiple Choice

Read each item carefully, and then select the best response.

_____ **1.** The term "state of matter" defines a substance as a(n):
A. solid, liquid, or gas
B. flammable or combustible
C. primary or secondary contaminant
D. alpha, beta, or gamma

_____ **2.** The first step in understanding the hazard of any chemical involves identifying:
A. physical properties
B. chemical properties
C. the state of matter
D. radiation agents

_____ **3.** The expansion ratio is a description of the volume increase that occurs when a material changes from:
A. a liquid to a solid
B. a solid to a gas
C. a solid to a liquid
D. a liquefied gas to a gas

_____ **4.** The ability of a chemical to undergo a change in its chemical make-up, usually with a release of some form of energy, is known as:
A. a property change
B. a physical change
C. chemical reactivity
D. a change of state

_____ **5.** Steel rusting and wood burning are examples of:
A. physical changes
B. chemical changes
C. vaporization
D. ionization

_____ **6.** The temperature at which a liquid will continually give off vapors, and will eventually turn completely into a gas, is known as the:
A. flash point
B. vaporization point
C. boiling point
D. gas point

_____ **7.** The weight of an airborne concentration, compared to an equal volume of dry air, is the:
A. vapor density
B. vapor ratio
C. flammable range
D. explosive ratio

_____ **8.** Air has a set vapor density value of:
A. 0.59
B. 1.0
C. 2.4
D. 14.7

_____ **9.** Corrosives are a complex group of chemicals that are categorized into which two classes?
A. simple and complex
B. primary and secondary
C. positive and negative pH
D. acid and base

_____ **10.** The ability of a substance to dissolve in water is known as its:
A. expansion ratio
B. persistence
C. water solubility
D. dispersion value

_____ **11.** pH is an expression of the concentration of:
A. hydrogen ions in a given substance
B. acid ions in a given substance
C. oxygen ions in a given substance
D. base ions in a given substance

_____ **12.** Common acids have pH values that are:
A. equal to zero
B. greater than 7
C. equal to 7
D. less than 7

_____ **13.** Bases have pH values that are:
A. equal to zero
B. greater than 7
C. equal to 7
D. less than 7

_____ **14.** The hazardous chemical compounds released when a material decomposes under heat are known as:
A. carcinogens
B. alpha particles
C. toxic products of combustion
D. beta particles

_____ **15.** The nucleus of a radioactive isotope includes an imbalance of the concentration of:
A. protons and neutrons
B. electrons and protons
C. electrons and neutrons
D. protons, neutrons, and electrons

_____ **16.** Of the following, which is the least penetrating type of radiation?
A. alpha
B. beta
C. gamma
D. neutron

_____ **17.** What is the process by which a person or object transfers contamination to another person or object through direct contact?
A. contamination by association
B. secondary exposure
C. direct contamination
D. secondary contamination

_____ **18.** What term describes when a single dose of a material causes death by any route other than inhalation?
A. lethal concentration
B. lethal dose
C. fatality level
D. toxic level

_____ **19.** Which type of chemical causes a substantial proportion of exposed people to develop an allergic reaction in normal tissue after repeated exposure?
A. sensitizer
B. irritant
C. convulsant
D. contaminant

_____ **20.** Adverse health effects caused by long-term exposure to a substance are known as:
A. acute health hazards
B. chronic health hazards
C. long-term disablers
D. overexposures

Labeling

Label the following diagrams with the correct terms.

1. Container breaches

Courtesy of Rob Schnepp.

A. ______________________________

Courtesy of Rob Schnepp.

B. ______________________________

Courtesy of Rob Schnepp.

C. ______________________________

Courtesy of Rob Schnepp.

D. ______________________________

2. Dispersion patterns

A. ____

B. ____

C. ____

D. ____

E

E. ______________________________

F. ______________________________

G. ______________________________

3. Vapor density

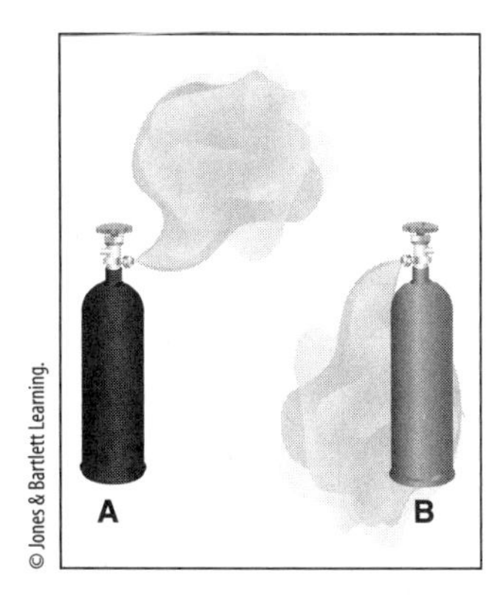

A. ______________________________

B. ______________________________

4. Types of radiation

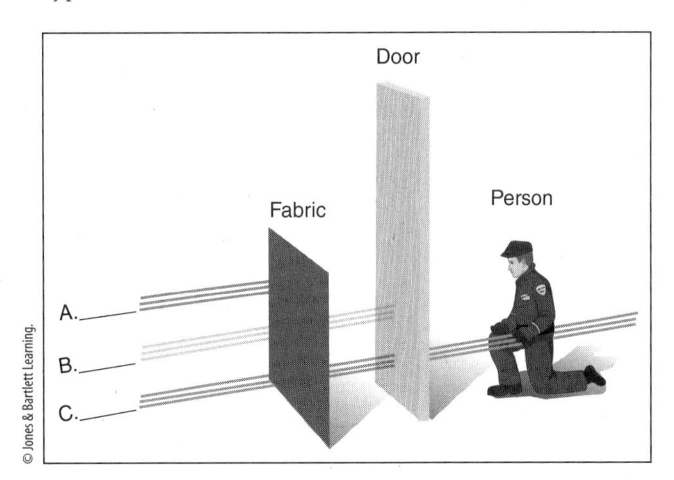

Vocabulary

Define the following terms using the space provided.

1. Ignition temperature

____

____

____

2. Expansion ratio

____

____

____

3. Radiation

____

____

____

4. Contamination

____

____

____

5. Specific gravity

____

____

____

Fill-In

Read each item carefully, and then complete the statement by filling in the missing word(s).

1. If the state of matter and physical properties of a chemical are known, a firefighter can ____________________ what the substance will do if it escapes its container.
2. Chemicals can undergo a(n) ____________________ change when subjected to outside influences such as heat, cold, and pressure.
3. Standard atmospheric pressure at sea level is ____________________ pounds per square inch.
4. The flash point of gasoline is ____________________.
5. The ____________________ ____________________ is the minimum temperature at which a liquid or solid emits vapor sufficient to form an ignitable mixture with air.
6. The ____________________ temperature is the minimum temperature at which a substance will ignite without an external ignition source.
7. Most flammable liquids will ____________________ on water.
8. The periodic table illustrates all the known ____________________ that make up every known compound.
9. ____________________ radiation particles can break chemical bonds creating ions; therefore, they are considered ionizing radiation.
10. Chemicals that pose a hazard to health after relatively short exposure periods are said to cause ____________________ ____________________ ____________________.

True/False

If you believe the statement to be more true than false, write the letter "T" in the space provided. If you believe the statement to be more false than true, write the letter "F."

1. _______ A chemical change occurs when the chemical is subjected to environmental influences such as heat, cold, or pressure.
2. _______ Water has an expansion rate of 100:1 and a boiling point of 100 degrees Fahrenheit (38 degrees Celsius).
3. _______ Diesel fuel has a higher flash point than gasoline.
4. _______ Wood burning is an example of polymerization.
5. _______ The wider the flammable range, the more dangerous the material.
6. _______ The nucleus of an atom is made up of protons and neutrons.
7. _______ Radioactive isotopes can be detected by the noise and odors they create.
8. _______ Liquids with low flash points typically have higher ignition temperatures.
9. _______ A bacteria or virus that can be transmitted from person to person is considered contagious.
10. _______ Alpha radiation is the most energetic type of radiation.

Short Answer

Complete this section with short written answers using the space provided.

1. List and explain the three (3) main types of stress that can cause a container to fail.

2. List five (5) of the items of information that would typically be found on the label on a pesticide bag.

3. Identify the two (2) factors that cause radiation to be a health hazard.

4. List five (5) of the notable substances found in most fire smoke.

5. What does "4H MEDIC ANNA" stand for?

Hazardous Material Alarms

The following real-case scenarios will give you an opportunity to explore the concerns associated with hazardous materials. Read each scenario, and then answer each question in detail.

1. You are dispatched to a local hospital for a hazardous materials incident. When your engine arrives at the scene, a lab technician states that she believes that one of the medical containers is leaking radioactive material. How should you proceed?

____

____

____

____

____

____

2. At 2:00 PM, your engine is dispatched to a local propane facility following the report of a "large propane tank on fire." Upon arrival, you find a small truck on fire and the flames are directly impinging a 500-pound propane tank. One of the many concerns you have is the possibility of a BLEVE occurring. Explain what a BLEVE is.

____

____

____

____

____

____

Understanding the Hazards

Workbook Activities

The following activities have been designed to help you. Your instructor may require you to complete some or all of these activities as a regular part of your hazardous materials training program. You are encouraged to complete any activity that your instructor does not assign you as a way to enhance your learning in the classroom.

Chapter Review

The following exercises provide an opportunity to refresh your knowledge of this chapter.

Matching

Match each of the terms in the left column to the appropriate definition in the right column.

______	**1.** Nerve agent	**A.**	Disease caused by a form of bacteria that is commonly found on rodents
______	**2.** Anthrax	**B.**	Toxic chemical agents that attack the central nervous system
______	**3.** Biological agent	**C.**	A medication that may reverse some of the effects of exposure to nerve agents
______	**4.** Smallpox	**D.**	An explosive device designed to injure emergency responders who have responded to an initial event
______	**5.** Plague	**E.**	Organisms that cause disease and attack the body
______	**6.** Secondary device	**F.**	Materials that emit radioactivity
______	**7.** Atropine	**G.**	Time period between the initial infection by an organism and the development of symptoms
______	**8.** Radiological agents	**H.**	A nerve agent that is primarily a vapor hazard
______	**9.** Sarin	**I.**	Highly infectious disease caused by a type of virus
______	**10.** Incubation period	**J.**	Infectious disease caused by a form of bacteria

Multiple Choice

Read each item carefully, and then select the best response.

______ **1.** Liquid bulk storage containers have an internal capacity of more than:

- **A.** 500 gallons
- **B.** 250 gallons
- **C.** 182 gallons
- **D.** 119 gallons

CHAPTER 4

_____ **2.** Solid bulk storage containers have an internal capacity of more than:
A. 1118 pounds
B. 1084 pounds
C. 919 pounds
D. 882 pounds

_____ **3.** One of the most common chemical tankers is a gasoline tanker, also known as a(n):
A. MC-331 pressure cargo tanker
B. MC-307/DOT 407 chemical tanker
C. MC-306/DOT 406 flammable liquid tanker
D. tube trailer

_____ **4.** Which type of container is generally V-shaped, with rounded sides, and is used to carry grain or fertilizers?
A. consist trailers
B. dry bulk cargo tankers
C. carboys
D. ASTs

_____ **5.** Firefighters should be able to recognize the three basic railcar configurations of:
A. non-pressurized, pressurized, special use
B. dry bulk, liquid, hazardous materials
C. agricultural, mechanical, commercial
D. contained, unpackaged, hazardous materials

_____ **6.** What term is used to describe the time period between the actual infection and the appearance of symptoms?
A. growth period
B. dispersing period
C. incubation period
D. implementation period

_____ **7.** A high-pressure MC-331 tank would carry what type of hazardous material?
A. diesel fuel
B. propane
C. poisons
D. cryogenics

_____ **8.** In the United States, the clearinghouse of technical chemical information is known as:
A. CHEMTREC
B. CANUTEC
C. CHEMRESPOND
D. SETIQ

______ **9.** Pressurized horizontal tanks usually have a small vapor space above the liquid level. This is known as the:
A. pressure space
B. liquid space
C. head space
D. reserve space

______ **10.** Intermodal tanks can be shipped by:
A. air
B. sea
C. land
D. all of the above

______ **11.** Secondary containment basins can typically hold:
A. half of the entire volume of the tank
B. the entire volume of the tank
C. the entire volume of the tank and 10% of anticipated rainfall
D. the entire volume of the tank and a percentage of the water from hose lines or sprinkler systems

______ **12.** The most common above-ground pressurized tanks contain:
A. gasoline
B. liquid propane
C. corrosives
D. oxidizers

______ **13.** Intermediate bulk containers are so named because the volume stored is typically between what is stored in drums or bags and what is stored in:
A. cargo tanks
B. ton containers
C. pressurized tanks
D. non-pressurized tanks

______ **14.** The transportation of hazardous materials most often occurs by:
A. sea
B. rail
C. road
D. air

______ **15.** A box-like structure attached to the rear of the tank and containing the tank controls is a feature of what type of cargo tank?
A. MC-306/DOT 406
B. MC-307/DOT 407
C. MC-331
D. MC-338

______ **16.** Small puffs of white vapor from the relief valve on a cryogenic tank should be considered a(n):
A. normal occurrence
B. indication that the relief valve is not operating properly
C. indication of a true emergency
D. indication that the tank is being used for non-cryogenic materials

______ **17.** Potential targets for terrorist activities can be classified into three broad categories that include infrastructure targets, civilian targets, and:
A. military targets
B. human-made targets
C. symbolic targets
D. transportation targets

______ **18.** Firefighters responding to a potential or known terrorist incident should use the same approach as they would for a(n):

A. structure fire incident
B. hazardous materials incident
C. EMS incident
D. technical rescue incident

______ **19.** Initial actions at a potential terrorist incident would include establishing Incident Command at what distance from the actual incident scene?

A. as far as 200 feet
B. between 300 and 500 feet
C. as far as 1000 feet
D. as far as 3000 feet

______ **20.** The most common type of clandestine laboratories encountered by responders are used to manufacture:

A. drugs
B. explosives
C. dirty bombs
D. biological agents

Labeling

Label the following diagrams with the correct terms.

Courtesy of Polar Tank Trailer L.L.C.

A. ______________________________

Courtesy of Polar Tank Trailer L.L.C.

B. ______________________________

Courtesy of National Tank Truck Carriers Association.

C. ______________________________

Courtesy of Rob Schnepp.

D. ______________________________

Courtesy of Jack B. Kelly, Inc.

E. ______________________________

Courtesy of Jack B. Kelly, Inc.

F. ______________________________

Courtesy of Polar Tank Trailer L.L.C.

G. ______________________________

Courtesy of private source.

H. ______________________________

Vocabulary

Define the following terms using the space provided.

1. Plague

____

____

____

2. Radiation dispersal device

____

____

____

3. Anthrax

____

____

____

4. Biological agents

____

____

____

Fill-In

Read each item carefully, and then complete the statement by filling in the missing word(s).

1. MC-__________ corrosives tankers are used for transporting sulfuric acid and other corrosive substances.
2. Compressed gases such as hydrogen, oxygen, and methane are carried by ______ trailers.
3. Flammable liquids, mild corrosives, and consumer products such as corn syrup are typically carried in ______________ rail tank cars.
4. ___________ liquid tank cars are the most common special-use railcars that responders may encounter.
5. Chlorine and ________ railcars have federal security alarms to prevent chemicals from being stolen or terrorist activities from occurring.
6. _________ _________ are toxic substances used to attack the central nervous system and were developed in Germany before World War II.
7. ___________ is a mnemonic used to remember the symptoms of possible nerve agent exposure.
8. The time period between the actual infection and the appearance of symptoms is known as the _________ __________.
9. The _______ ________ ________ has established reporting requirements for different chemicals based on the reportable quantity (RQ) for that chemical.
10. Indicators of potential secondary devices such as timers, wires, or switches are known as ______ ______.

True/False

If you believe the statement to be more true than false, write the letter "T" in the space provided. If you believe the statement to be more false than true, write the letter "F."

1. ______ Covered floating roof tanks are commonly used to store combustible liquids.
2. ______ The U.S. Department of Transportation does not consider tube trailers to be cargo tanks.
3. ______ Chemical cargo tanks are often identifiable by the presence of several heavy-duty reinforcing rings around the tank.
4. ______ Tube trailers will only carry one type of product.

5. _______ Tank cars that carry chlorine are an example of special-use railcars.
6. _______ Atropine can be used to reverse some of the effects of nerve agent exposure.
7. _______ The biggest difference between a chemical incident and a biological incident is typically the speed of onset of the health effects from the involved agents.
8. _______ Diseases such as smallpox and pneumonic plague cannot be passed from person to person.
9. _______ Food-testing labs are an example of a facility that routinely uses radioactive materials.
10. _______ While smallpox is highly contagious, it only has a mortality rate of approximately 10 percent.

Short Answer

Complete this section with short written answers using the space provided.

1. What are some of the factors regarding potential terrorism targets that responders in any size community must consider?

____

____

____

____

2. What are some of the things that responders should be aware of regarding "dirty bombs"?

____

____

____

____

Hazardous Material Alarms

The following real-case scenarios will give you an opportunity to explore the concerns associated with hazardous materials. Read each scenario, and then answer each question in detail.

1. When responding to a known or potential terrorist incident, what are some of the initial actions that should be considered by responders?

____

____

____

____

____

____

2. Your engine company is part of a multiple community response to an incident that potentially involves a large group of people being exposed to nerve agents. Upon arrival, you are assigned to assist with triage and emergency medical care. What are some of the signs and symptoms you should be prepared to see?

____

____

____

____

____

____

Estimating Potential Harm and Planning a Response

Workbook Activities

The following activities have been designed to help you. Your instructor may require you to complete some or all of these activities as a regular part of your hazardous materials training program. You are encouraged to complete any activity that your instructor does not assign you as a way to enhance your learning in the classroom.

Chapter Review

The following exercises provide an opportunity to refresh your knowledge of this chapter.

Matching

Match each of the terms in the left column to the appropriate definition in the right column.

_____ **1.** TLV/C — **A.** The maximum concentration of hazardous material that a worker should be exposed to, even for an instant

_____ **2.** Defensive objectives — **B.** The process by which a substance moves through a given material on the molecular level

_____ **3.** TLV/TWA — **C.** The maximum airborne concentration of a material to which a worker can be exposed for 8 hours a day, 40 hours a week, and not suffer any ill effects

_____ **4.** IDLH — **D.** A value established by NIOSH that is comparable to OSHA's PEL

_____ **5.** REL — **E.** Describing an atmospheric concentration of a toxic, corrosive, or asphyxiant substance that poses an immediate threat to life or could cause irreversible or delayed adverse health effects

_____ **6.** Sheltering-in-place — **F.** The removal or relocation of individuals who may be affected by an approaching release of a hazardous material

_____ **7.** Evacuation — **G.** The physical destruction or decomposition of a clothing material due to chemical exposure, general use, or ambient conditions

_____ **8.** Penetration — **H.** A method of safeguarding people by temporarily keeping them in a cleaner atmosphere

_____ **9.** Permeation — **I.** Examples include diking and damming, diluting or diverting material, or suppressing or dispersing vapor

_____ **10.** Degradation — **J.** The flow or movement of a hazardous material through closures

Multiple Choice

Read each item carefully, and then select the best response.

_____ **1.** When planning an initial hazardous materials incident response, what is the first priority?

A. Consider the effect on the environment.
B. Consider the safety of the victims.
C. Consider the equipment and personnel needed to mitigate the incident.
D. Consider the safety of the responding personnel.

CHAPTER 5

_______ **2.** Planning a response begins with the:
A. size-up
B. initial call for help
C. Incident Commander's orders
D. review of standard operating procedures

_______ **3.** Responders to a hazardous materials incident need to know the:
A. type of material involved
B. general operating guidelines
C. short- and long-term effects of the hazardous material
D. duration of the incident

_______ **4.** Selection of personal protective equipment is based on the:
A. hazardous materials involved
B. level of training of the responder
C. direction of the Incident Commander
D. standard operating procedures of the department

_______ **5.** One of the primary objectives of a medical surveillance program is to determine:
A. the intensity of the response at an incident
B. the concentration of the chemicals at the incident
C. the time of duty at an incident
D. any changes in the functioning of body systems

_______ **6.** Responders to hazardous materials incidents need to consider:
A. the size of the container
B. the nature and amount of the material released
C. the area exposed to the material
D. all of the above

_______ **7.** Litmus paper is used to determine:
A. the time at which contamination occurred
B. pH
C. weather
D. the location of the contamination

_______ **8.** Defensive actions include:
A. plugging
B. patching
C. diking
D. overpacking

_______ **9.** When performing isolation activities, the specific type will be determined by the nature of the released chemical and:
A. available staffing
B. environmental conditions
C. the potential number of victims
D. the incident action plan

______ **10.** When the hazard of entering the hot zone is too great, this is known as:

A. nonintervention
B. shelter in place
C. denial of entry
D. isolation

Labeling

Label the following diagrams with the correct terms.

1. Level ____________________ ensemble

2. Level ____________________ ensemble

3. Level ____________________ ensemble

4. Level ____________________ ensemble

Vocabulary

Define the following terms using the space provided.

1. Defensive objectives

____

____

____

2. Isolation of the hazard zone

____

____

____

3. Decontamination corridor

____

____

____

4. Chemical-resistant materials

____

____

____

5. Supplied-air respirator (SAR)

____

____

____

Fill-In

Read each item carefully, and then complete the statement by filling in the missing word(s).

1. When a hazardous material incident is detected, there should be an initial call for more ______________.

2. ______________ burns are often much deeper and more destructive than acid burns.

3. Two main organizations establish and publish the toxicological data used by hazardous materials responders. They are the American Conference of Governmental Industrial Hygienists (ACGIH) and ______________.

4. ______________ ______________ is enhanced by abrasions, cuts, heat, and moisture.

5. The process of transferring a hazardous material from its source to people, animals, the environment, or equipment is known as ______________ contamination.

6. ______________ paper can be used to determine the concentration of an acid or a base by reporting the hazardous material's pH.

7. ______________ ______________ protective equipment is a level above structural firefighter gear and shields the wearer during short-term exposures to high temperatures.

8. Diking and damming, stopping the flow of a substance remotely, and diluting or directing a material are examples of ______________ actions.

9. ______________ ______________ ______________ is a method of safeguarding people located near or in a hazardous area by temporarily keeping them in a cleaner atmosphere, usually inside structures.

10. ______________ is the process by which a hazardous chemical moves through a given material to the molecular level.

True/False

If you believe the statement to be more true than false, write the letter "T" in the space provided. If you believe the statement to be more false than true, write the letter "F."

1. ______ Degradation and penetration refer to the same process.
2. ______ The Emergency Response Guidebook (ERG) is a resource for determining evacuation distances.
3. ______ A Level B ensemble provides a high level of respiratory protection but less skin protection than Level A.
4. ______ A Level D ensemble is the highest level of protection.
5. ______ Air-purifying respirators (APRs) provide breathing air and are appropriate for use when sufficient oxygen for breathing is not available.
6. ______ The rate of absorption of a hazardous chemical can vary depending on the body part involved.
7. ______ Firefighter structural firefighting gear is considered the first level of chemical-protective ensembles.
8. ______ The selection and use of chemical protective clothing may have the greatest direct impact on responder health and safety.
9. ______ In regard to corrosive liquids, the concentration is an expression of how much acid or base is dissolved in a solution, usually water.
10. ______ The standard unit of measure for estimating the extent of a hazardous materials release is square feet.

Short Answer

Complete this section with short written answers using the space provided.

1. Identify ten (10) pieces of information that could be reported to agencies to in their preparation for a response to a hazardous materials incident.

____

____

____

____

2. Identify and define the three basic atmospheres at a hazardous materials emergency according to the exposure guidelines.

____

____

____

____

3. Identify and provide a brief description of the four levels of protective clothing.

____

____

____

____

Hazardous Material Alarms

The following real-case scenarios will give you an opportunity to explore the concerns associated with hazardous materials. Read each scenario, and then answer each question in detail.

1. It's 4:40 in the afternoon when your engine company is dispatched to a local industrial plant to assist with a hazardous materials spill. Upon arrival, the Incident Commander states that several employees were splashed with an orange-colored liquid and that you need to set up an emergency decontamination area. What actions will you take to accomplish this?

____

____

____

____

____

____

2. It's 2:30 in the afternoon when your engine company is dispatched to a four-story hotel where several people have been affected by chlorine vapors from the indoor swimming pool area. Upon arrival, the Incident Commander tells you that a shelter-in-place approach will be used to isolate individuals who are on the upper floors of the hotel. Your assignment is to assist with accomplishing this. What are some of your considerations in using this method?

____

____

____

____

____

____

Implementing the Planned Response

Workbook Activities

The following activities have been designed to help you. Your instructor may require you to complete some or all of these activities as a regular part of your hazardous materials training program. You are encouraged to complete any activity that your instructor does not assign you as a way to enhance your learning in the classroom.

Chapter Review

The following exercises provide an opportunity to refresh your knowledge of this chapter.

Matching

Match each of the terms in the left column to the appropriate definition in the right column.

______	**1.** Cold zone	**A.**	The safe area that houses the command post at an incident
______	**2.** Operations	**B.**	The point of contact for the media
______	**3.** Public Information Officer	**C.**	Used where multiple agencies with overlapping jurisdictions or responsibilities are involved in the same incident
______	**4.** Hot zone	**D.**	The section of the command structure that carries out the objectives developed by the IC
______	**5.** Warm zone	**E.**	Refers to crews and companies working in the same geographic location
______	**6.** Unified Command	**F.**	Gathers information and reports to the Incident Commander and the Safety Officer
______	**7.** Liaison officer	**G.**	The point of contact for cooperating agencies on scene
______	**8.** Technical Reference Team	**H.**	The starting point for implementing any response
______	**9.** Size-up	**I.**	The area immediately surrounding a hazardous materials incident
______	**10.** Division	**J.**	The decontamination corridor is located here

Multiple Choice

Read each item carefully, and then select the best response.

______ **1.** To determine how far to extend evacuation distances, hazardous materials technicians should:

- **A.** contact product specialists
- **B.** refer to the ERG
- **C.** get direction from the Incident Commander
- **D.** use detection and monitoring equipment

______ **2.** Which of the following factors is a major concern when considering evacuation?

- **A.** estimated duration of the release
- **B.** distance to a safe area
- **C.** time of day
- **D.** amount of property involved in the incident

CHAPTER

6

______ **3.** Which method of safeguarding people in a hazardous area involves keeping them in a safe atmosphere?

A. staying indoors
B. duck and cover
C. shelter in place
D. isolation zone

______ **4.** During the initial size-up at a hazardous materials incident, the first decision concerns:

A. the amount of property affected
B. responder safety
C. the number of people involved
D. the type of material involved

______ **5.** Evacuation distances for small spills or fires involving hazardous materials are listed in the:

A. blue pages of the ERG
B. green pages of the ERG
C. orange pages of the ERG
D. yellow pages of the ERG

______ **6.** The first step in gaining control of a hazardous materials incident is to isolate the problem and:

A. equip the cold zone
B. keep people away
C. establish a backup team
D. identify the hazardous materials involved

______ **7.** Designated areas at a hazardous materials incident based on safety and the degree of hazard are called:

A. control zones
B. hot zones
C. isolation zones
D. control corridors

______ **8.** What is the area immediately surrounding the incident known as?

A. control zone
B. hot zone
C. warm zone
D. cold zone

______ **9.** What is the area where personnel and equipment are staged before they enter and after they leave the hot zone?

A. control zone
B. hot zone
C. warm zone
D. cold zone

______ **10.** What is the safe area in which personnel do not need to wear any special protective clothing for safe operation?

A. control zone
B. hot zone
C. warm zone
D. cold zone

Labeling

Label the following diagrams with the correct terms.

1. The DECIDE Model

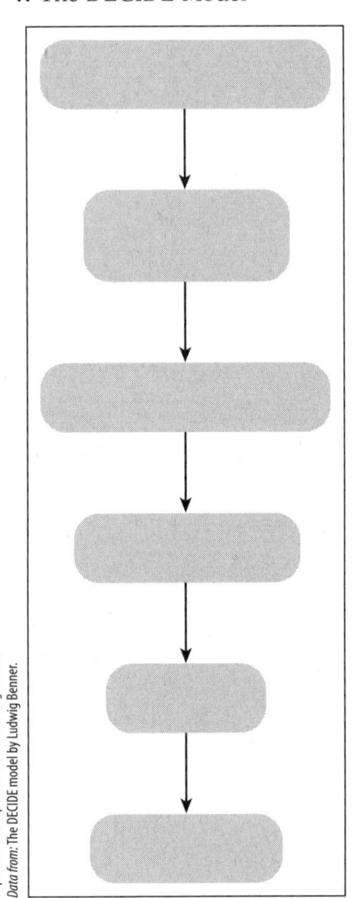

2. Control zones

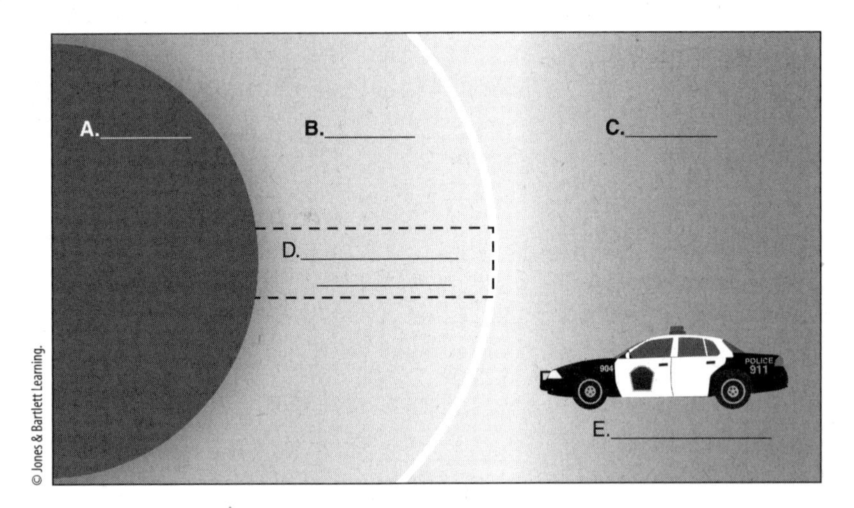

Vocabulary

Define the following terms using the space provided.

1. Shelter in place

 ____

 ____

 ____

2. Safety briefing

 ____

 ____

 ____

3. Heat exhaustion

 ____

 ____

 ____

4. Heat stroke

 ____

 ____

 ____

5. Logistics section

 ____

 ____

 ____

Fill-In

Read each item carefully, and then complete the statement by filling in the missing word(s).

1. The protection of ________ is the first priority in an emergency response situation.
2. Before an evacuation order is given, a(n) ________ ________ and suitable shelter are established.
3. The age, health, and ________ of evacuees are key factors to be considered in the decision to begin an evacuation.
4. The individual in charge of the ICS shall designate a(n) ________ ________ who has a specific responsibility to identify and evaluate hazards.
5. The simplest expression of the ________ ________ is for no fewer than two responders to enter a contaminated area.
6. Isolating the scene, protecting the exposures, and allowing the incident to stabilize on its own are examples of ________ actions.
7. Wet clothing extracts heat from the body as much as ________ times faster than dry clothing.

8. ________ is the rapid mental process of evaluating the critical visual indicators of the incident, processing that information, and arriving at a conclusion that formulates your plan of action.

9. The layer of clothing next to the skin, especially the ________, should always be kept dry.

10. Reference sources such as the ________ can be useful in formulating response plans.

True/False

If you believe the statement to be more true than false, write the letter "T" in the space provided. If you believe the statement to be more false than true, write the letter "F."

1. ______ The duration of the hazardous materials incident is a factor in determining whether shelter in place is a viable option.
2. ______ Monitoring and portable detection devices assist the Incident Commander in determining the hot, warm, and cold zones, as well as the evacuation distances required.
3. ______ The safety of responders is paramount to maintaining an effective response to any hazardous materials incident.
4. ______ In most hazardous materials incidents, the first action that should be taken is to begin evacuation.
5. ______ Backup personnel remain on standby in the cold zone, awaiting orders to prepare for follow-up duties.
6. ______ The decontamination team reports to the Logistics section of ICS.
7. ______ There are several ways to isolate the hazard area and create the control zones.
8. ______ The warm zone contains control points for access corridors, as well as the decontamination corridor.
9. ______ All personnel must be fully briefed before they approach the hazard area or enter the cold zone.
10. ______ An incident that involves a gaseous contaminant will require a larger hot zone than one involving a liquid leak.

Short Answer

Complete this section with short written answers using the space provided.

1. Identify and describe some of the key benefits of the Incident Command System.

____

____

____

____

2. Identify and describe the four major functional components within the Incident Command System.

____

____

____

____

3. Identify and provide a brief description of the three zones at a hazardous materials incident.

____

____

____

____

4. Identify and provide a brief description of the positions that comprise the command staff as part of the Incident Command System.

____

____

____

____

Hazardous Material Alarms

The following real-case scenarios will give you an opportunity to explore the concerns associated with hazardous materials. Read each scenario, and then answer each question in detail.

1. It is 3:00 AM on a Saturday when your engine company responds to a hazardous materials incident on a local highway. Upon arrival, the Incident Commander directs you to supervise the establishment of the control zones. What are some of your considerations for accomplishing this?

____

____

____

____

____

____

2. It is 2:00 PM on a Tuesday and you respond to an off-duty callback for a significant hazardous materials incident at a local manufacturing plant. Upon arrival, the Fire Chief assigns you to assist with setting up the Incident Command post. What are some of the factors you should consider in accomplishing this?

____

____

____

____

____

____

Responder Health and Safety

Workbook Activities

The following activities have been designed to help you. Your instructor may require you to complete some or all of these activities as a regular part of your hazardous materials training program. You are encouraged to complete any activity that your instructor does not assign you as a way to enhance your learning in the classroom.

Chapter Review

The following exercises provide an opportunity to refresh your knowledge of this chapter.

Matching

Match each of the terms in the left column to the appropriate definition in the right column.

______	**1.** Hypoxia	**A.** A group of substances that are found in fire smoke
______	**2.** Pyrolysis	**B.** A colorless and odorless gas created during incomplete combustion
______	**3.** PAHs	**C.** A sensor that uses ultraviolet light
______	**4.** Carbon monoxide	**D.** A state of inadequate oxygenation
______	**5.** Amyl nitrate	**E.** A collection of particulates usually called smoke and that is a human carcinogen
______	**6.** Photo ionization sensor	**F.** A process in which material is decomposed or broken down
______	**7.** Electrochemical sensor	**G.** A sensor that uses a chemical reagent
______	**8.** Soot	**H.** A highly toxic gas found in fire smoke
______	**9.** Cyanide	**I.** One of the medications used to treat cyanide poisoning
______	**10.** Oxygen	**J.** Considered the natural antidote for CO poisoning

Multiple Choice

Read each item carefully, and then select the best response.

______ **1.** Smoke production depends on the temperature of the combustion process, the chemical make-up of the burning material, and the:

A. influence of ventilation
B. type of structure
C. distance from the seat of the fire
D. amount of time the fire has been burning

CHAPTER 7

______ **2.** The most predominant substance in a typical mattress is:
A. laminates
B. nylon
C. polyurethane foam
D. rubber

______ **3.** The Immediately Dangerous to Life and Health (IDLH) level of carbon monoxide is:
A. 35 ppm
B. 500 ppm
C. 1200 ppm
D. 1700 ppm

______ **4.** A PID sensor uses ____________ to ionize the gases that move through the sensor.
A. ethane
B. oxygen
C. radioactive material
D. ultraviolet light

______ **5.** The United States is placed within the top ______ countries in the world for per capita fire deaths.
A. 10
B. 15
C. 50
D. 100

______ **6.** Headache, nausea, dizziness, and fatigue are all common signs and symptoms of:
A. heat exhaustion
B. overexertion
C. chemical overdose
D. breathing smoke

______ **7.** Volume, color, and the ____ of smoke are all good indicators of what the fire is doing inside the structure.
A. force
B. location
C. stratification
D. heaviness

______ **8.** Recent studies dispel a long-held belief that carbon monoxide is the predominant killer in fire smoke. In fact, it appears that ________ plays a role in smoke-related death and injury.
A. ammonia
B. benzene
C. cyanide
D. pyrolysis

_______ **9.** Until recently, there have been three primary uses for detection devices, outside traditional fire-based hazardous materials response. These uses include CO detector responses; building collapse and trench rescue; and rescue response, including _________.

A. water rescue
B. confined space rescue
C. industrial entrapment rescue
D. high-angle rescue

_______ **10.** _______________ establishes the safe levels of chemical exposures in the workplace.

A. OSHA
B. NFPA
C. United States EPA
D. NIOSH

Vocabulary

Define the following terms using the space provided.

1. Hypoxia

____

____

____

2. Pyrolysis

____

____

____

3. Anaerobic metabolism

____

____

____

4. Aerobic metabolism

____

____

____

Fill-In

Read each item carefully, and then complete the statement by filling in the missing word(s).

1. ____________________ is one of the first observable signs of a working fire.
2. Poisoning from ____________________ may be more predominant than carbon monoxide poisoning as a cause of death in certain fire victims.
3. ____________________ is the natural antidote for CO poisoning.
4. Synthetic substances ignite and burn ____________________ causing developing fires and toxic smoke.
5. When PAHs are generated during a fire, they may bind with ____________________, resulting in dermal and inhalation exposure.
6. Carbon monoxide affects the oxygen-carrying capacity of the ____________________ ____________________ ____________________, resulting in hypoxia.
7. While CO reduces the amount of oxygen carried to the cells, ____________________ renders the cells incapable of using whatever oxygen is present.
8. Exposure levels are based on the exposure of an average worker and are ____________________ for firefighters.
9. The Recommended Exposure Limit (REL) for carbon monoxide is ____________________ ppm.
10. Carbon monoxide sensors are filled with a jelly-like substance containing ____________________ ____________________ and two electrical poles within the sensor.

True/False

If you believe the statement to be more true than false, write the letter "T" in the space provided. If you believe the statement to be more false than true, write the letter "F."

1. ______ Carbon monoxide has a distinctive odor.
2. ______ Soot is a known human carcinogen.
3. ______ After a fire is knocked down, it is common to see firefighters wearing SCBA during overhaul.
4. ______ Cancer is more prevalent in the fire service than in most other occupations.
5. ______ Carbon monoxide is one of the most common industrial hazards.
6. ______ Since 2004, there has been a nationwide standard protocol for treating smoke inhalation.
7. ______ In the United States, there are currently two types of approved cyanide poisoning interventions.
8. ______ The NFPA has developed a best practice for detection and monitoring of the fire environment.
9. ______ There is no single device that can detect all common fire gases.
10. ______ Electrochemical sensors should be bump tested prior to use.

Short Answer

Complete this section with short written answers using the space provided.

1. Identify what carbon monoxide (CO) is and why it is dangerous.

2. Identify five (5) of the signs and symptoms of carbon monoxide exposure.

3. Identify some of the actions that can be taken by firefighters to reduce the harm from smoke exposure.

4. Identify and briefly describe the three (3) primary types of exposure levels.

Hazardous Material Alarms

The following real-case scenario will give you an opportunity to explore the concerns associated with hazardous materials. Read the scenario, and then answer the question in detail.

1. You are a newly promoted officer in your department. Your shift commander has asked you to prepare a short training class to remind personnel that it's not just those on the interior of a structure fire that are exposed to smoke. What are some of the items you should include in your lesson plan?

____

____

____

____

____

____

Operations Mission-Specific: Personal Protective Equipment

Workbook Activities

The following activities have been designed to help you. Your instructor may require you to complete some or all of these activities as a regular part of your hazardous materials training program. You are encouraged to complete any activity that your instructor does not assign you as a way to enhance your learning in the classroom.

Chapter Review

The following exercises provide an opportunity to refresh your knowledge of this chapter.

Matching

Match each of the terms in the left column to the appropriate definition in the right column.

______	**1.** Degradation	**A.**	Offers the lowest level of protection
______	**2.** Dehydration	**B.**	The flow or movement of a hazardous chemical through closures such as zippers, seams, or imperfections in material
______	**3.** Level D ensemble	**C.**	Usually precedes heat exhaustion, heat cramps, and heat stroke
______	**4.** Permeation	**D.**	The physical destruction of clothing material as a result of chemical exposure
______	**5.** Penetration	**E.**	Respiratory protection sometimes worn with a Level C ensemble
______	**6.** APRs	**F.**	A process that is similar to water saturating a sponge
______	**7.** SARs	**G.**	Worn when the airborne substance is known, criteria for APR are met, and skin and eye exposure is unlikely
______	**8.** Level A ensemble	**H.**	Respiratory protection sometimes worn with a Level A ensemble
______	**9.** Level C ensemble	**I.**	Used when the hazardous material identified requires the highest level of protection for skin, eyes, and respiration
______	**10.** NIOSH	**J.**	Sets the design, testing, and certification standards for SCBA in the United States

Multiple Choice

Read each item carefully, and then select the best response.

______ **1.** Time, distance, and shielding are the preferred methods of protection for which of the following?

- **A.** biological exposures
- **B.** chemical exposures
- **C.** radiation exposures
- **D.** high thermal exposures

______ **2.** The process by which a hazardous chemical moves through closures, seams, or porous materials is called:
A. penetration
B. degradation
C. permeation
D. vaporization

______ **3.** The physical destruction of clothing due to chemical exposure is called:
A. penetration
B. degradation
C. permeation
D. vaporization

______ **4.** Chemical resistance, flexibility, abrasion, temperature resistance, shelf life, and sizing criteria are requirements that need to be considered when selecting:
A. monitoring equipment
B. respirators
C. testing equipment
D. chemical-protective clothing

______ **5.** Vapor-protective clothing requires the wearer to use:
A. air-purifying respirators
B. powered air-purifying respirators
C. particulate respirators
D. self-contained breathing apparatus

______ **6.** All of the following may have an impact on the service life of chemical-resistant materials, EXCEPT ONE. CHOOSE THE EXCEPTION:
A. permeation
B. degradation
C. penetration
D. sublimation

______ **7.** Which protective ensemble should be worn by responders entering a poorly ventilated area with an ammonia leak?
A. vapor-protective clothing
B. liquid splash–protective clothing
C. high-temperature–protective equipment
D. structural fire protective clothing

______ **8.** The *Standard on Vapor-Protective Ensembles for Hazardous Materials Emergencies* has been established by:
A. OSHA
B. NFPA
C. EPA
D. NIOSH

_______ **9.** Which is one of the two properties on which protective clothing chemical compatibility charts are based?
- **A.** breakthrough time
- **B.** flame resistance
- **C.** tensile strength
- **D.** thermal reflectivity

_______ **10.** The organs and muscles of the torso generate about _____% of all body heat.
- **A.** 50
- **B.** 65
- **C.** 75
- **D.** 90

_______ **11.** Which is the most common type of illness caused by wearing PPE?
- **A.** respiratory
- **B.** abrasions
- **C.** heat-related
- **D.** mechanical injury

Labeling

Label the following diagrams with the correct terms.

1. Types of personal protective equipment:

A. ____

© Photodisc.

B. ______________________________

© Jones & Bartlett Learning. Photographed by Glen E. Ellman.

C. ______________________________

2. Common hand signals:

Courtesy of Rob Schnepp.

A. ______________________________

Courtesy of Rob Schnepp.

B. ______________________________

Courtesy of Rob Schnepp.

C. ______________________________

Vocabulary

Define the following terms using the space provided.

1. Level A ensemble

____

____

____

2. Dehydration

____

____

____

3. Donning

____

____

____

4. High-temperature–protective equipment

____

____

____

5. Powered air-purifying respirator (PAPR)

____

____

____

Fill-In

Read each item carefully, and then complete the statement by filling in the missing word(s).

1. Work uniforms offer the _______ amount of protection in a hazardous materials emergency.
2. _______________ protective clothing is designed to prevent chemicals from coming in contact with the body and may have varying degrees of resistance.
3. A fully encapsulated suit is a(n) _______________-piece garment that completely encloses the wearer.
4. A gel-packed vest is an example of a(n) _______________.
5. _______________ is an acronym used to sum up a collection of potential hazards that an emergency responder may face.
6. Firefighters should be encouraged to drink approximately _______________ of water before donning any protective clothing.

7. Levels A, B, C, and D ensemble classifications have been established by EPA and OSHA ______________________________ regulations.

8. Current OSHA regulations require __________ to ensure that the PPE worn at a hazardous materials emergency is appropriate for the hazards encountered.

9. There is a well-defined relationship between __________ and heat emergencies.

10. Phase-change cooling technology operates in a similar fashion to __________.

True/False

If you believe the statement to be more true than false, write the letter "T" in the space provided. If you believe the statement to be more false than true, write the letter "F."

1. ______ Standard firefighting turnout gear offers little chemical protection, but it does have a high degree of abrasion resistance and prevents direct skin contact.

2. ______ Tyvek provides satisfactory protection from all chemicals.

3. ______ Vapor-protective clothing and chemical-protective clothing are identical.

4. ______ Degradation is the process by which a chemical moves through a given material on the molecular level.

5. ______ Reusable Level A suits are required to be pressure-tested after each use.

6. ______ Single-use chemical-protective clothing is expected to be discarded, along with the hazardous waste generated by the incident.

7. ______ The rapid and destructive way that gasoline dissolves a Styrofoam cup is an example of degradation.

8. ______ *Breakthrough time* is the time it takes a substance to be absorbed into the suit fabric and detected on the other side.

9. ______ A proximity suit combines a high degree of thermal and chemical protection.

10. ______ OSHA HAZWOPER regulations require the use of the buddy system.

Short Answer

Complete this section with short written answers using the space provided.

1. What seven hazards does the acronym TRACEMP refer to?

____

____

____

____

2. Identify and provide a brief description of the four levels of protective clothing.

____

____

____

____

Skill Drills

Test your knowledge of this skill drill by placing the following photos in the correct order. Number the first step with a "1," the second step with a "2," and so on.

Skill Drill 8-1: Donning a Level A Ensemble NFPA 1072: 6.2.1

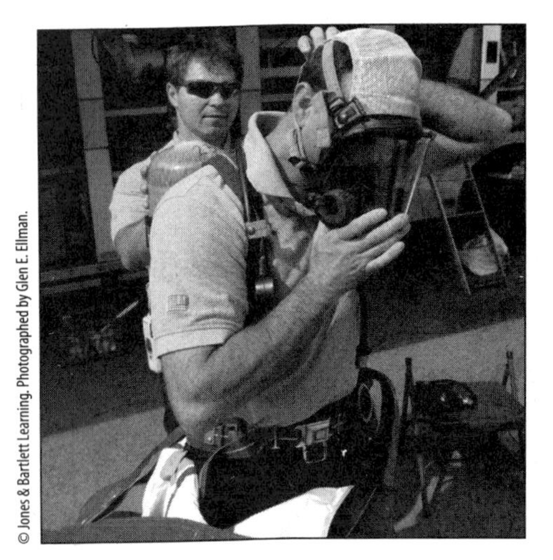

______ Stand up and don the SCBA frame and SCBA face piece, but do not connect the regulator to the face piece.

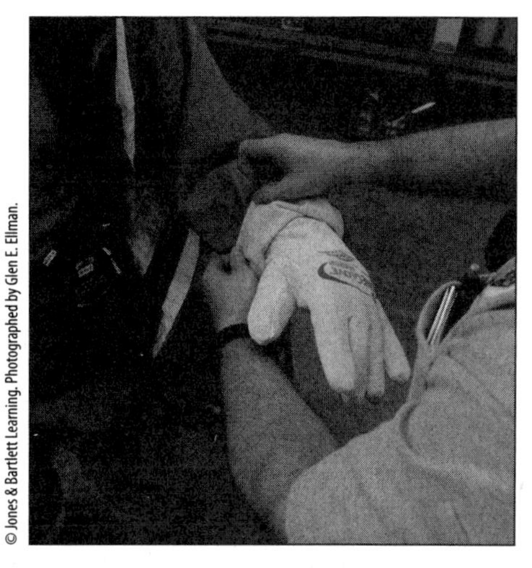

______ Don the outer chemical gloves (if required). With assistance, complete donning the suit by placing both arms in the suit, pulling the expanded back piece over the SCBA, and placing the chemical suit over your head.

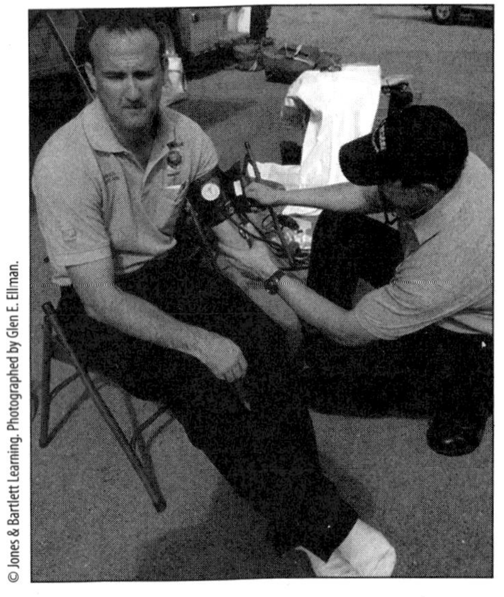

______ Conduct a pre-entry briefing, medical monitoring, and equipment inspection.

______ Review hand signals, and indicate that you are okay.

3. What does CBRN stand for?

Hazardous Materials Alarms

The following real-case scenarios will give you an opportunity to explore the concerns associated with hazardous materials. Rea each scenario, and then answer each question in detail.

1. During hazardous materials response training, you are assigned to an entry team wearing Level B nonencapsulated personal protective clothing. What is the recommended PPE for Level B protection?

2. After 45 minutes of training while wearing your Level B PPE, you begin feeling poorly.

   a. What is the probable cause of your sudden illness?

   b. What are two possible preventative measures for this problem?

______ Instruct the assistant to connect the regulator to the SCBA face piece and ensure air flow.

______ Place the helmet on your head.

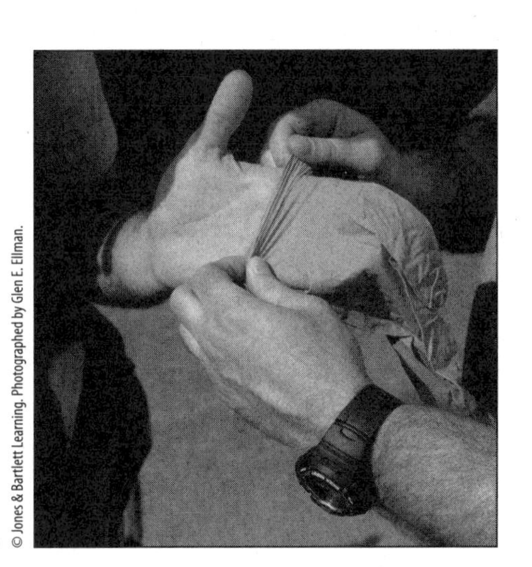

______ Don the inner gloves.

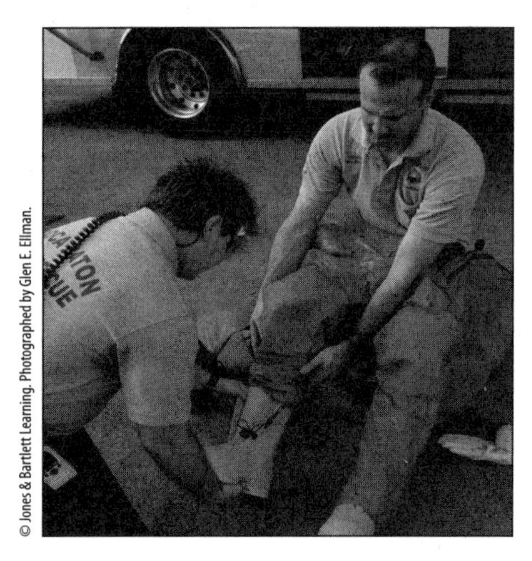

______ While seated, pull on the suit to waist level; pull on the chemical boots over the top of the chemical suit. Fold the suit boot covers over the tops of the boots.

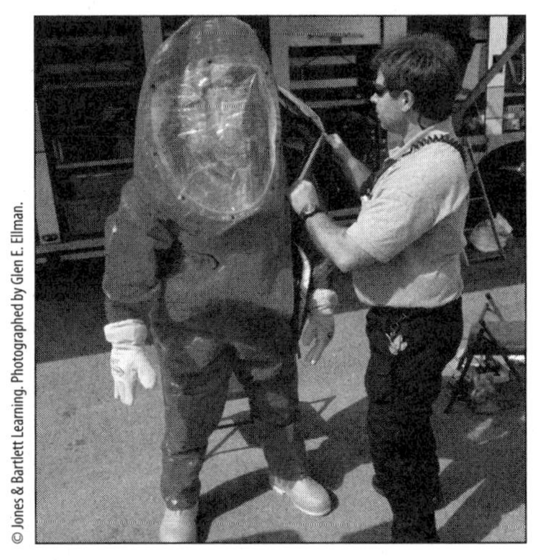

______ Instruct the assistant to close the chemical suit by closing the zipper and sealing the splash flap.

Skill Drill 8-2: Doffing a Level A Ensemble NFPA 1072: 6.2.1

Test your knowledge of this skill drill by filling in the correct words in the following photo captions.

1. After completing decontamination, proceed to the __________ __________ for suit doffing. Pull your hands out of the outer gloves and your arms from the sleeves, and cross your arms in front inside the suit.
2. Instruct the assistant to open the __________ __________ flap and suit zipper.
3. Instruct the assistant to begin at the head and roll the suit __________ and __________ until the suit is below waist level.
4. Instruct the assistant to complete rolling the suit from the waist to the __________; step out of the attached chemical boots and suit.
5. Doff the SCBA frame. The ____ ______ should be kept in place while the SCBA frame is doffed.
6. Take a deep breath and doff the SCBA face piece; carefully peel off the inner gloves, and then walk away from the clean area. Go to the rehabilitation area for __________ __________, rehydration, and a personal decontamination shower.

Skill Drill 8-3: Donning a Level B Nonencapsulated Chemical-Protective Clothing Ensemble NFPA 1072: 6.2.1

Test your knowledge of this skill drill by placing the following photos in the correct order. Number the first step with a "1", the second step with a "2", and so on.

______ Sit down, and pull on the suit to waist level; pull on the chemical boots over the top of the chemical suit.

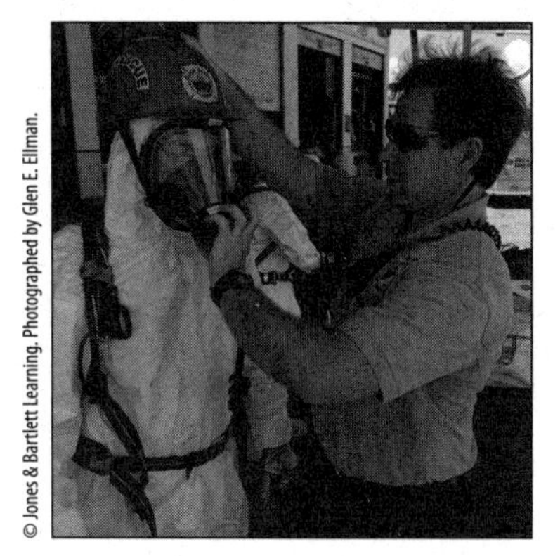

______ With assistance, pull the hood over your head and the SCBA face piece. Place the helmet on your head. Put on the outer gloves (over or under the sleeves, depending on the AHJ requirements for the incident). Instruct the assistant to connect the regulator to the SCBA face piece, and ensure you have air flow. Review hand signals, and indicate that you are okay.

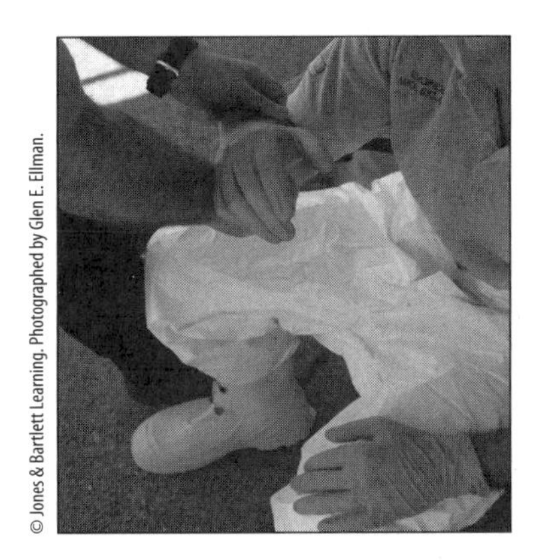

______ Don the inner gloves.

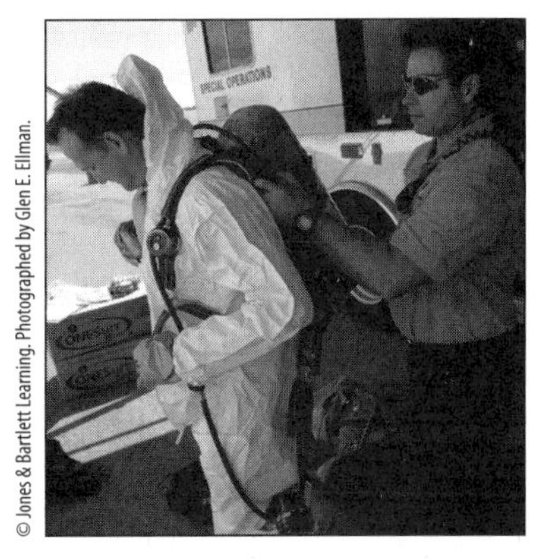

______ Don the SCBA frame and SCBA face piece, but do not connect the regulator to the face piece.

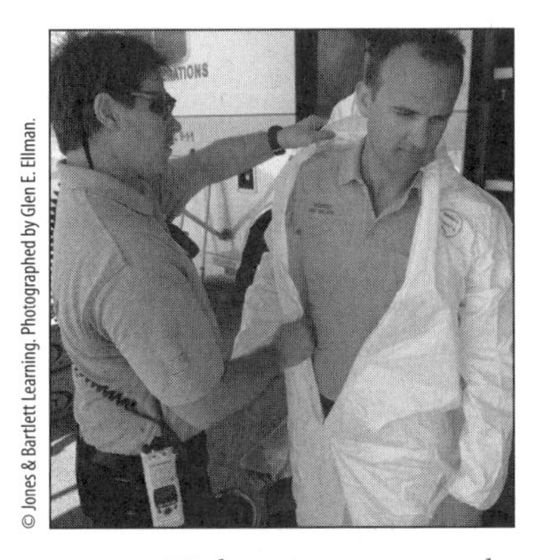

______ With assistance, complete donning the suit by placing both arms in the suit and pulling the suit over your shoulders. Instruct the assistant to close the chemical suit by closing the zipper and sealing the splash flap.

______ Conduct a pre-entry briefing, medical monitoring, and equipment inspection.

Skill Drill 8-4: Doffing a Level B Nonencapsulated Chemical-Protective Clothing Ensemble NFPA 1072: 6.2.1

Test your knowledge of this skill drill by filling in the correct words in the photo captions.

1. After completing the ______________/______________ cycle, proceed to the clean area for PPE doffing. The SCBA frame is removed first. The unit may remain attached to the regulator while the assistant helps the responder out of the PPE, or the air supply may be detached from the regulator, leaving the face piece in place to provide for face and eye protection while the rest of the doffing process is completed.
2. Instruct the assistant to open the ______________ splash flap and suit zipper.
3. Remove your hands from the outer gloves and your arms from the sleeves of the suit. ____________ ____________ ______________ in front, inside the suit. Instruct the assistant to begin at the head and roll the suit down and away until the suit is below waist level.
4. Sit down and instruct the assistant to complete rolling down the suit to the ankles; step out of the attached chemical boots and ________.
5. Stand and doff the SCBA face piece and ________.
6. Carefully peel off the inner gloves, and then go to the rehabilitation area for medical monitoring, rehydration, and ______________ ______________ ______________.

Skill Drill 8-5: Donning a Level C Chemical-Protective Clothing Ensemble NFPA 1072: 6.2.1

Test your knowledge of this skill drill by placing the following photos in the correct order. Number the first step with a "1," the second step with a "2," and so on.

______ Don the APR/PAPR face piece. With assistance, pull the hood over your head and the PAR/ PAPR face piece. Place the helmet on your head. Pull on the outer gloves. Review hand signals, and indicate that you are okay.

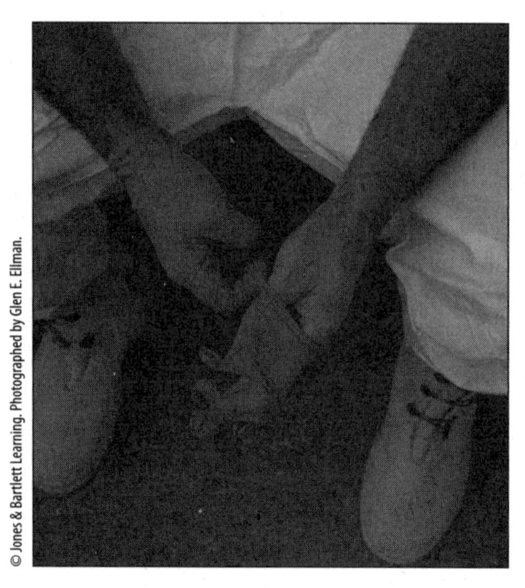

______ Don the inner gloves.

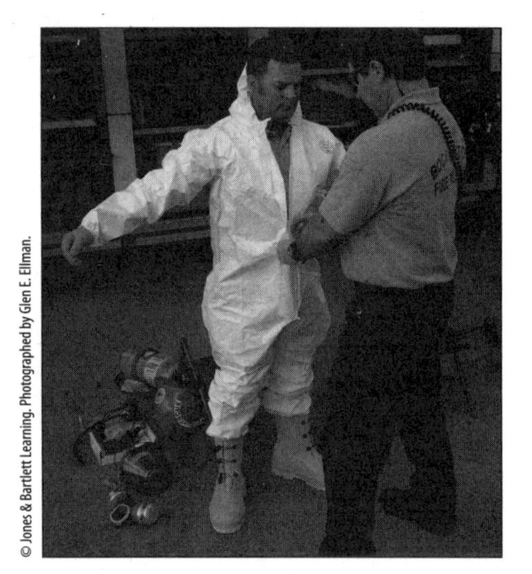

______ With assistance, complete donning the suit by placing both arms in the suit and pulling the suit over your shoulders. Instruct the assistant to close the chemical suit by closing the zipper and sealing the splash flap.

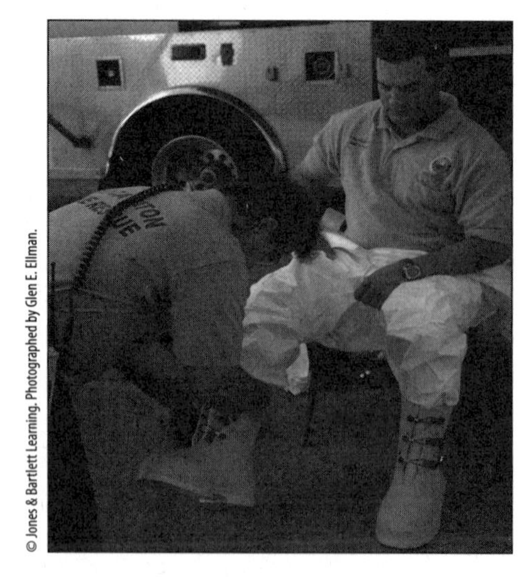

______ Conduct a pre-entry briefing, medical monitoring, and equipment inspection. While seated, pull on the suit to waist level; pull on the chemical boots over the top of the chemical suit.

Skill Drill 8-6: Doffing a Level C Chemical-Protective Clothing Ensemble NFPA 1072: 6.2.1

Test your knowledge of this skill drill by filling in the correct words in the photo captions.

1. After completing decontamination, proceed to the clean area. As with Level B, the assistant opens the chemical splash flap and suit zipper. Remove your hands from the gloves and your arms from the sleeves. Instruct the assistant to begin at the head and roll the suit down below waist level. Instruct the assistant to complete rolling down the suit and to take the outer boots and suit away. The assistant helps remove the inner gloves. Remove the ________________/________________. Remove the helmet.
2. Go to the rehabilitation area for medical monitoring, ________________, and a personal decontamination shower.

Skill Drill 8-7: Donning a Level D Chemical-Protective Clothing Ensemble NFPA 1072: 6.2.1

Test your knowledge of this skill drill by filling in the correct words in the photo caption.

1. Conduct a pre-entry briefing, medical monitoring, and equipment inspection. Don the Level D suit. Don boots. Don ____________ ____________ or ________________ ________________. Don a hard hat. Don gloves, a face shield, and any other required equipment.

Operations Mission-Specific: Technical Decontamination

Workbook Activities

The following activities have been designed to help you. Your instructor may require you to complete some or all of these activities as a regular part of your hazardous materials training program. You are encouraged to complete any activity that your instructor does not assign you as a way to enhance your learning in the classroom.

Chapter Review

The following exercises provide an opportunity to refresh your knowledge of this chapter.

Matching

Match each of the terms in the left column to the appropriate definition in the right column.

	Term	Definition
_____	**1.** Decontamination	**A.** The process used to destroy disease carrying microorganisms
_____	**2.** Chemical degradation	**B.** The process of mixing a spongy material into a spilled liquid and collecting the contaminated mixture
_____	**3.** Sterilization	**C.** The process of chemically treating a hazardous liquid to turn it into a solid material, making the material easier to handle
_____	**4.** Absorption	**D.** A natural or artificial process that causes the breakdown of a chemical substance
_____	**5.** Solidification	**E.** The process of adding a material to a contaminant, which then adheres to the surface of the material for collection
_____	**6.** Disinfection	**F.** The physical or chemical process of reducing and preventing the spread of contaminants to people, animals, and the environment.
_____	**7.** Adsorption	**G.** A process using heat, chemical means, or radiation to kill microorganisms
_____	**8.** Dilution	**H.** The removal of dusts, particles, and some liquids by sucking them up into a container
_____	**9.** Evaporation	**I.** Commonly uses plain water to lower the concentration of a hazardous material while flushing it off a contaminated person or object
_____	**10.** Vacuuming	**J.** A natural form of chemical degradation that allows a chemical substance to stabilize without human intervention

CHAPTER 9

Multiple Choice

Read each item carefully, and then select the best response.

______ **1.** What is the process of transferring a hazardous material, or the hazardous component of a weapon of mass destruction, from its source to people, animals, the environment, or equipment?
A. contamination
B. dispersion
C. transference
D. integration

______ **2.** Which agency is responsible for laws and regulations governing the disposal of absorbent materials?
A. fire department
B. federal government
C. Department of Transportation
D. municipal/county government

______ **3.** Which method of decontamination is used on large groups of people with the goal of removing contaminants as quickly as possible?
A. emergency decontamination
B. group decontamination
C. gross decontamination
D. mass decontamination

______ **4.** Which decontamination procedure mixes a spongy material with a liquid hazardous material?
A. absorption
B. adsorption
C. dilution
D. vapor dispersal

______ **5.** Which is a removal process for items that cannot be properly decontaminated?
A. disinfection
B. solidification
C. isolation and disposal
D. rapid mass decontamination

______ **6.** During decontamination, what is usually the last item of clothing removed?
A. shoes
B. SCBA mask
C. inner gloves
D. face shield

______ **7.** Removed equipment should be placed:
A. near the entrance to the decontamination corridor
B. in the hot zone
C. in the cold zone
D. at the equipment cache

______ **8.** After full completion of the decontamination process, personnel should:
- **A.** report to their immediate supervisor
- **B.** be medically evaluated
- **C.** check in with the accountability officer
- **D.** report to the staging area

______ **9.** Which method is used to reduce the corrosivity of an acid or base?
- **A.** saponification
- **B.** emulsification
- **C.** reduction
- **D.** neutralization

______ **10.** Which vacuum system filter should be used when vacuuming hazardous materials?
- **A.** N95
- **B.** ionic
- **C.** HEPA
- **D.** charcoal

Labeling

Label the following diagrams with the correct terms.

Physical methods of technical decontamination.

1.

____

____

2. 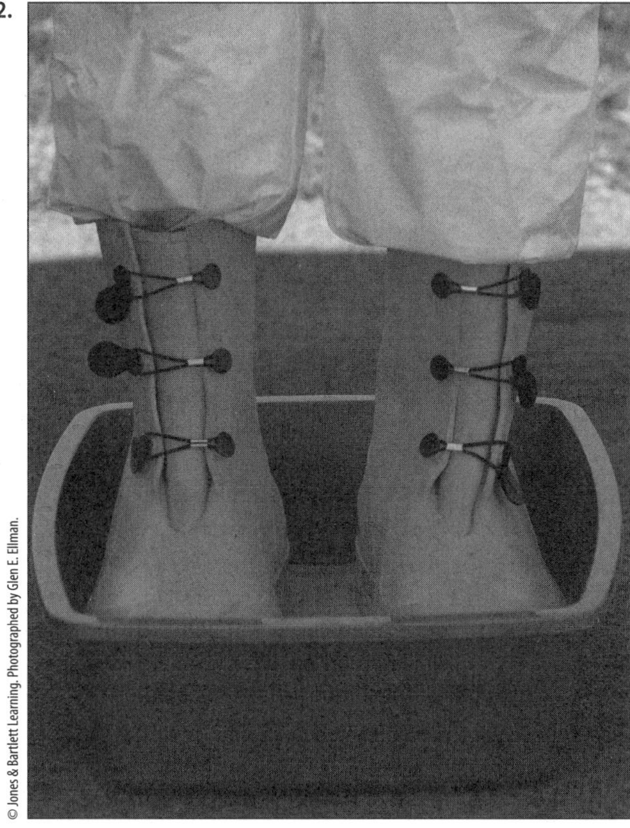

Vocabulary

Define the following terms using the space provided.

1. Decontamination team

2. Contamination

3. Adsorption

4. Solidification

____

____

____

5. Sterilization

____

____

____

Fill-In

Read each item carefully, and then complete the statement by filling in the missing word(s).

1. The ____________ ____________ is a controlled area, usually within the warm zone, where decontamination procedures take place.

2. ____________ ____________ consists of a pre-wash that occurs before technical decontamination can take place.

3. When volatile liquids are spilled, responders may elect to take no action and instead allow the substance to ____________.

4. The process other than sterilization that is used to destroy disease-causing microorganisms, excluding spores, is called ____________.

5. Most gross and mass decontamination processes use the decontamination technique of ____________.

6. ____________ involves removing contaminated items from the primary incident site and storing them in a designated area; ____________ is the legal transportation of these items to approved facilities to be stored, incinerated, buried in a hazardous waste landfill, or otherwise handled.

7. Sand is an example of a material commonly used in the decontamination process of ____________.

8. ____________ ____________ is performed after gross decontamination and is a more thorough cleaning process.

9. Prior to any contaminated responders or victims passing through the decontamination corridor, the corridor can be considered ____________.

10. Whenever possible, ____________ of the hazardous material before beginning decontamination.

True/False

If you believe the statement to be more true than false, write the letter "T" in the space provided. If you believe the statement to be more false than true, write the letter "F".

1. ______ Emergency medical responders are responsible for establishing a decontamination corridor for the initial emergency response crews and victims.

2. ______ During gross decontamination, hospital staff use low-pressure, high-volume water flow to rinse off and dilute contaminants.

3. _______ Vacuuming is the removal of dusts, particles, and some liquids by sucking them into a container.
4. _______ Personnel leaving the hot zone should place used tools in a tool drop area near the decontamination corridor.
5. _______ It is not possible to maintain a strict chain of custody for items of evidence that require contamination.

Short Answer

Complete this section with a short written answer using the space provided.

1. Identify and provide a brief description of the four major categories of decontamination.

____

____

____

____

Hazardous Material Alarms

The following real-case scenarios will give you an opportunity to explore the concerns associated with hazardous materials. Read each scenario, and then answer each question in detail.

1. Your engine company is dispatched to a pesticide spill at a local hardware store. On arrival, you see several store patrons covered with liquid and powder. Your officer orders your company to set up emergency decontamination for the contaminated patrons. How will you proceed?

____

____

____

____

____

____

2. Your Lieutenant has given you the chance to prepare a short presentation on alternative decontamination procedures. Which topics will you discuss?

____

____

____

____

____

____

Skill Drills

Skill Drill 9-1: Performing Technical Decontamination on a Responder or go through as a reponder NFPA 1072: 6.2.1, 6.4.1

Test your knowledge of this skill drill by placing the following photos in the correct order. Number the first step with a "1", the second step with a "2", and so on.

______ For technical decontamination, the decontamination team member washes and rinses the contaminated responder one to three times. The wash–rinse cycle is determined largely by the nature of the contaminant. Remember, the goal is to render the PPE safe to remove.

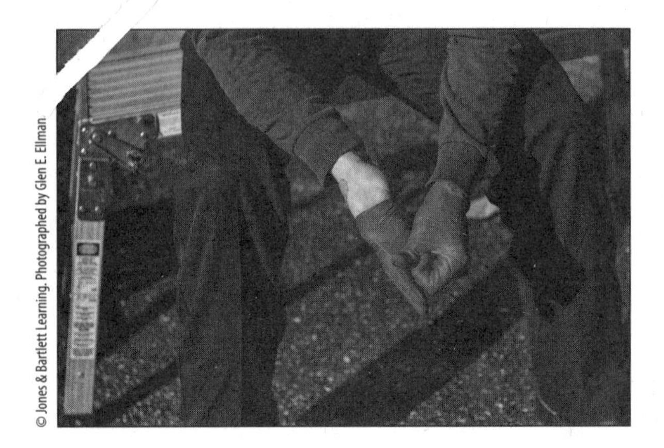

______ The responder removes his or her personal clothing and any respiratory protection, and proceeds to the rehabilitation area for medical monitoring, rehydration, and personal decontamination.

______ The contaminated responder drops any tools or equipment into a container or onto a designated tarp.

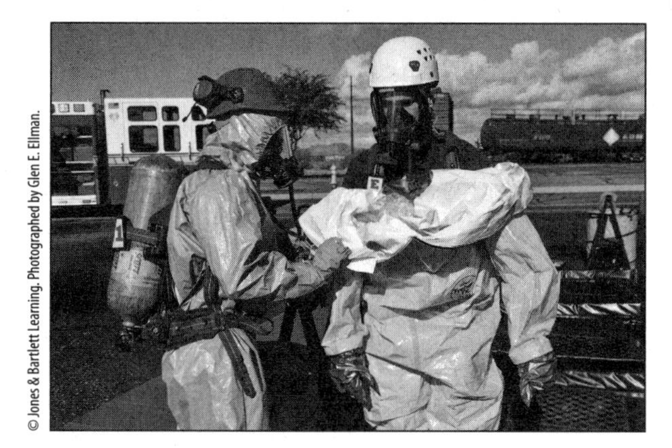

______ The decontamination team member removes the outer hazardous materials–protective clothing from the contaminated responder.

______ The decontamination team member performs gross decontamination on the contaminated responder, if necessary.

Operations Mission-Specific: Mass Decontamination

Workbook Activities

The following activities have been designed to help you. Your instructor may require you to complete some or all of these activities as a regular part of your hazardous materials training program. You are encouraged to complete any activity that your instructor does not assign you as a way to enhance your learning in the classroom.

Chapter Review

The following exercises provide an opportunity to refresh your knowledge of this chapter.

Matching

Match each of the terms in the left column to the appropriate definition in the right column.

	Term	Definition
______	**1.** Ambulatory victims	**A.** The process of adding some substance—usually water—to a contaminant to decrease its concentration
______	**2.** Washing	**B.** A process for removing contaminated items that cannot be properly decontaminated from the incident scene
______	**3.** Dilution	**C.** Those victims who are able to walk
______	**4.** Isolation and disposal	**D.** Dousing victims with a simple soap-and-water solution and then fully rinsing with water
______	**5.** ERG	**E.** Used as a quick reference for responders operating at a hazardous materials incident

Multiple Choice

Read each item carefully, and then select the best response.

______ **1.** All of the following are essential to the success of a mass decontamination operation EXCEPT ONE. CHOOSE THE EXCEPTION:
- **A.** Identify the contaminant if at all possible.
- **B.** Select and use the proper level of personal protective equipment.
- **C.** Have a predetermined process or procedure to perform decontamination.
- **D.** Assess the amount of property affected.

______ **2.** All of the following are acceptable methods of mass decontamination EXCEPT ONE. CHOOSE THE EXCEPTION:
- **A.** fog-type nozzles attached to pumpers opposite each other
- **B.** aerial ladder device providing an overhead spray pattern
- **C.** small volume straight streams
- **D.** prepackaged mass decontamination showers

______ **3.** Which is a fundamental difference between mass decontamination and other forms of decontamination?
A. the types of contaminants involved
B. the setting in which it occurs
C. the numbers of victims involved
D. the size of the control zones

______ **4.** Which technique is particularly suitable for dealing with contaminated personal clothing?
A. dilution
B. isolation
C. washing
D. disposal

______ **5.** Which group is the primary audience for the ERG?
A. allied professionals
B. first responders
C. medical practitioners
D. technician-level personnel

______ **6.** Never perform chemical neutralization of an acid or base if:
A. it is a water-reactive material
B. the material is water-soluble
C. the material is on human skin
D. there are nearby combustible exposures

______ **7.** A straight-water wash would be most effective with which material?
A. acid
B. nerve agent
C. benzene
D. waste oil

Labeling

Label the following diagrams with the correct terms.

1. An example of a simple mass decontamination corridor using two fire engines.

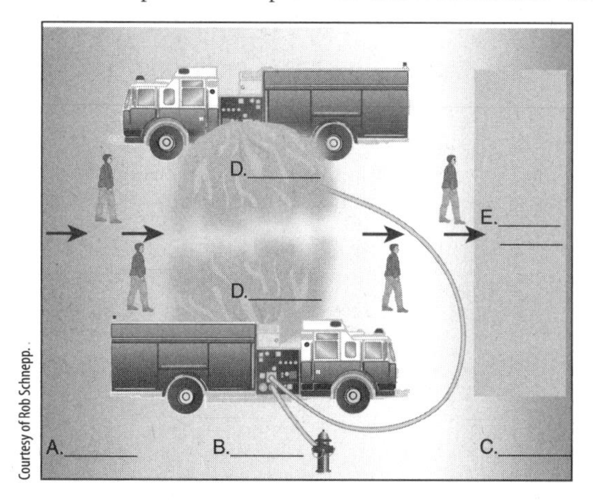

Courtesy of Rob Schnepp.

Vocabulary

Define the following terms using the space provided.

1. Mass decontamination

____

____

____

2. Dilution

____

____

____

3. Isolation and disposal

____

____

____

Fill-In

Read each item carefully, and then complete the statement by filling in the missing word(s).

1. A water temperature of ________________ is ideal but may not be possible.
2. In most cases, significant medical treatment should be provided after ____________________.
3. When flushing nonambulatory victims with water, take care to avoid compromising the victim's ____________________ with water during the process.
4. In mass decontamination, there are __________ processes used to reduce or eliminate contamination.
5. When possible, make every attempt to track the __________ and ________________ taken from the victims.

True/False

If you believe the statement to be more true than false, write the letter "T" in the space provided. If you believe the statement to be more false than true, write the letter "F."

1. ______ Mass decontamination and emergency decontamination are similar, except that emergency decontamination needs to be addressed much more quickly.
2. ______ Water washing will not change the physical or chemical properties of CO, so the application of water would yield minimum benefit.
3. ______ Controlling runoff is a primary objective when mass decontamination is implemented.
4. ______ Decontaminating ambulatory victims is a much slower process than decontaminating nonambulatory victims.
5. ______ When decontaminating nonambulatory victims, be sure to leave clothing or towels underneath the victim to allow for absorption of the product.

6. ______ Because water is a good general-purpose solvent, washing off as much of the contaminant as possible with a massive water spray is the best and quickest way to decontaminate a large group of people.
7. ______ Contaminated patients will almost always wait for responders to establish a formal decontamination area.
8. ______ Some viscous chemicals cannot be completely removed from the skin by washing alone.
9. ______ Evidence preservation is not an important consideration in mass decontamination activities.
10. ______ Naturally occurring barriers can be used to direct a moving group and create a manageable traffic flow.

Short Answer

Complete this section with short written answers using the space provided.

1. Describe the locations where the NFPA 704 system might be used.

____

____

____

____

2. Describe the locations where a placard system might be used.

____

____

____

____

Hazardous Material Alarms

The following real-case scenarios will give you an opportunity to explore the concerns associated with hazardous materials. Read each scenario, and then answer each question in detail.

1. Your agency has dispatched two engines, an ambulance, and a command vehicle to a university indoor sports arena where a large number of attendees have been exposed to an unknown agent that is severely irritating their skin. On arrival, you estimate that there are hundreds of anxious people with painful reddening of their skin streaming out of the main entrance. Your nearest mobile decontamination equipment will be coming from a neighboring town with a 20-minute ETA. How would you perform mass decontamination in this situation?

____

____

____

____

____

____

2. University officials are expressing concern about the runoff from the mass decontamination process. Are they correct in their concern?

____

____

____

____

____

____

Skill Drills

Skill Drill 10-1: Performing Mass Decontamination on Ambulatory Victims NFPA 1072: 6.3.1

Test your knowledge of this skill drill by filling in the correct words in the photo captions.

1. Ensure that you have the appropriate ______________ to protect yourself from the chemical threat, and do not make physical contact with it. Direct victims out of the hazard zone to a suitable location.
2. Set up the appropriate type of mass decontamination system based on the nature of the contaminant and the type of ______________, equipment, and/or system available.
3. Instruct victims to remove all contaminated clothing and walk through the decontamination process. ______________ the contaminated victims with water.
4. Direct the contaminated victims to a(n) ______________/______________ evaluation area.

Skill Drill 10-2: Performing Mass Decontamination on Nonambulatory Victims NFPA 1072: 6.3.1

Test your knowledge of this skill drill by placing the following photos in the correct order. Number the first step with a "1", the second step with a "2", and so on.

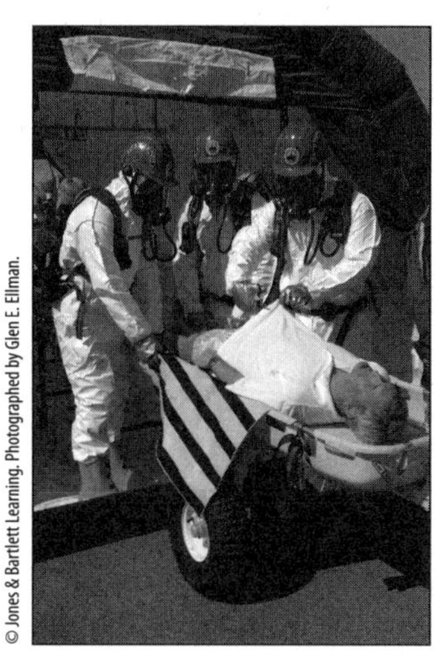

______ Ensure you have the appropriate PPE to protect against the chemical threat. Remove the appropriate amount of the victim's clothing. Do not leave any clothing underneath the victim; these items may wick the contamination to the victim's back and hold it there, potentially worsening the exposure.

______ Flush the contaminated victims with water. Make sure to rinse well under and around the straps that may be holding the victim to a backboard or other extrication device. *Take care to avoid compromising the victim's airway with water during the process.*

______ Set up the appropriate type of mass decontamination system based on the nature of the contaminant, type of apparatus, equipment, and/or decontamination system available.

______ Move the victims through the decontamination corridor and into the triage area for medical evaluation.

Operations Mission-Specific: Evidence Preservation and Sampling

Workbook Activities

The following activities have been designed to help you. Your instructor may require you to complete some or all of these activities as a regular part of your hazardous materials training program. You are encouraged to complete any activity that your instructor does not assign you as a way to enhance your learning in the classroom.

Chapter Review

The following exercises provide an opportunity to refresh your knowledge of this chapter.

Matching

Match each of the terms in the left column to the appropriate definition in the right column.

	Term	Definition
_____	**1.** Evidence	**A.** A minute quantity of physical evidence that is conveyed from one place to another
_____	**2.** Investigative authority	**B.** Refers to all of the information that is gathered and used by an investigator in determining the cause of an incident
_____	**3.** Chain of custody	**C.** The agency that has the legal jurisdiction to enforce a local, state, or federal law or regulation and is the appropriate law enforcement organization to investigate and prosecute
_____	**4.** Trace evidence	**D.** Anything that can be used to validate a theory to show how something could have occurred
_____	**5.** Demonstrative evidence	**E.** The process of maintaining continuous possession and control of the evidence from the time it is discovered until the time it is presented in court

Multiple Choice

Read each item carefully, and then select the best response.

_____ **1.** Which agency is responsible for investigating suspicious letters or packages involving hazardous materials that are sent through the mail?

- **A.** Postal Inspection Service
- **B.** Drug Enforcement Administration
- **C.** Federal Bureau of Investigation
- **D.** Environmental Protection Agency

_____ **2.** A suspect's clothing that has residue of the same ignitable liquid found at the scene of a fire is an example of ________________ evidence.

- **A.** physical
- **B.** residual
- **C.** trace
- **D.** demonstrative

______ **3.** Burn patterns on a wall or an empty gasoline can left at the scene of an incident are examples of ____________________ evidence.

A. analytical
B. residual
C. direct
D. physical

______ **4.** Information that can be used to prove a theory based on facts that were observed firsthand is considered to be ____________________ evidence

A. circumstantial
B. theoretical
C. investigative
D. deductive

______ **5.** Fences, excessive window coverings, or enhanced ventilation and air filtration systems may be indicators of:

A. evidence collection facilities
B. illicit laboratories
C. EOD facilities
D. direct evidence

______ **6.** Evidence used in the legal process to establish a fact or prove a point is often referred to as ____________________ evidence.

A. forensic
B. documentary
C. procedural
D. demonstrative

______ **7.** Which characteristic of a hazardous materials incident frequently makes the use of unified command necessary?

A. difficulty in maintaining a manageable span of control
B. the need to preserve a strict chain of custody
C. the likelihood of subsequent legal action
D. the involvement of multiple agencies

Vocabulary

Define the following terms using the space provided.

1. Chain of custody

____

____

____

2. Investigative authority

____

____

____

3. Demonstrative evidence

____

____

____

Fill-In

Read each item carefully, and then complete the statement by filling in the missing word(s).

1. ____________________ evidence consists of items that can be observed, photographed, measured, collected, examined in a laboratory, and presented in court to prove or demonstrate a point.
2. Facts that can be observed or reported firsthand are referred to as ____________________ evidence.
3. The process of protecting potential evidence until it can be documented, sampled, and collected appropriately is referred to as ____________________ ____________________.
4. The first step in developing an evidence preservation plan is identifying a method to ____________________.
5. It is essential that, whenever an explosive device is suspected of being involved, the appropriate ____________________ should be notified.
6. ____________________/____________________ records kept by the Incident Commander are a good way to document the identity and purpose of all personnel who enter an area that has been categorized as a crime scene.
7. Biological agents should be packaged only in ____________________ and ____________________ containers to ensure that cross-contamination with other microorganisms does not occur.
8. The use of a computer model to demonstrate how a fire could spread would be an example of ____________________ evidence.
9. Hazardous materials incidents are often very complex and dynamic situations. It is not uncommon for a ____________________ ____________________, consisting of multiple disciplines and jurisdictional agencies, to be established at such events.
10. ____________________ are often the first link in the chain of custody.

True/False

If you believe the statement to be more true than false, write the letter "T" in the space provided. If you believe the statement to be more false than true, write the letter "F."

1. ______ Regardless of the type of attack or weapon dissemination method, it is imperative that evidence be preserved, sampled, and collected properly.
2. ______ An intentional release or attack involving hazardous materials would be investigated by the Environmental Protection Agency.
3. ______ Even in incidents where no crime has been committed, evidence recovery may be essential.

4. ______ Hazardous materials first responders at the scene are in the best position to decide whether the evidence they find will be admissible in court and worthy of preservation.

5. ______ Plastic containers should not be used to hold evidence-containing petroleum products because these chemicals may lead to deterioration of the plastic.

Short Answer

Complete this section with short written answers using the space provided.

1. Describe the 12-step process, recommended by the FBI, regarding the collection or sampling of evidence.

____

____

____

____

2. List three indicators that legitimate toxic industrial chemicals were released intentionally.

____

____

____

____

Hazardous Material Alarms

The following real case scenarios will give you an opportunity to explore the concerns associated with hazardous materials. Read each scenario, and then answer each question in detail.

1. You have been dispatched to a fire alarm in a residential community. On your arrival, a police officer indicates that there is no fire, but he states that neighbors suspect a clandestine laboratory is operating there. What indicators should you look for to consider this risk?

____

____

____

____

____

____

2. Your size-up of the area gives you the impression that the operators of this illegal laboratory may also have been disposing of hazardous wastes in the back yard of the property. What indicators should you look for to consider this risk?

____

____

____

____

____

____

Skill Drills

Skill Drill 11-1: Collecting Samples and Preserving Evidence NFPA 1072: 6.5.1

Test your knowledge of this skill drill by placing the following photos in the correct order. Number the first step with a "1", the second step with a "2", and so on.

______ Sketch, mark, and label the location of the sample. Sketch the scene as near to scale as possible.

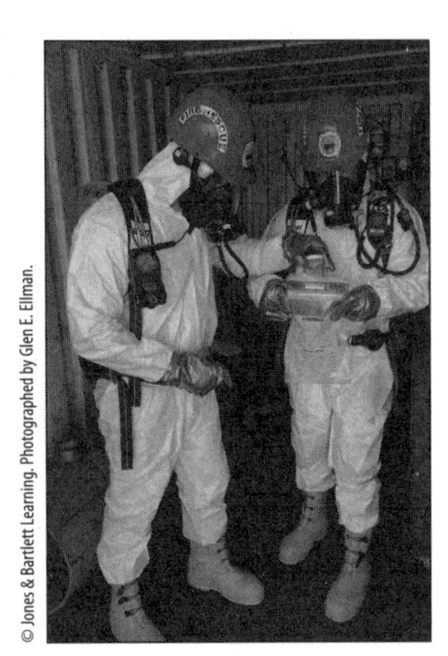

______ Label the sample. Anything being transported for analysis should include a label with the date, time, location, discoverer's name, and witnesses' names.

______ Take photographs of each sample as it is found and collected. If possible, photograph the item exactly as it was found, before it is moved or disturbed.

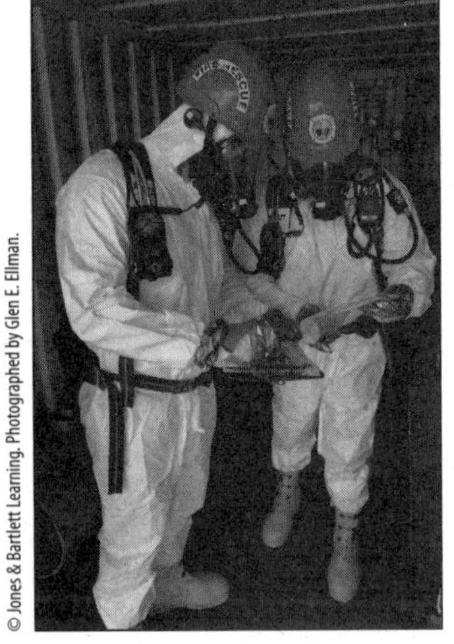

______ Record the time when the evidence/sample was discovered, the location where it was found, and the name of the person who found it.

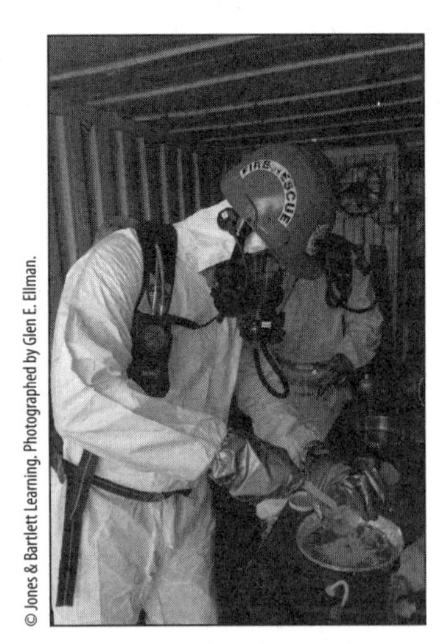

______ Place the sample in an appropriate container to ensure its safety and prevent contamination.

Skill Drill 11-2: Securing, Characterizing, and Preserving the Scene NFPA 1072: 6.5.1

Test your knowledge of this skill drill by filling in the correct words in the photo captions.

1. Observe the scene for certain characteristics that could lead to the discovery of evidence. Assess the number of victims and property damage, if any, and the type (such as a liquid or a solid) and quantity of __________ materials on site.
2. Secure the scene by placing __________ or __________ __________ to limit access to the scene.
3. Preserve suspected evidence by protecting it from being __________.

Skill Drill 11-3: Documenting the Activity of Personnel NFPA 1072: 6.5.1

Test your knowledge of this skill drill by filling in the correct words in the photo caption.

1. Keep a written log or record of the name and __________ of each responder who enters the scene.

Skill Drill 11-4: Implementing Response Actions NFPA 1072: 6.5.1

Test your knowledge of this skill drill by filling in the correct words in the photo captions.

1. Identify the presence of suspected __________ materials or devices.
2. Establish a safe perimeter, and consider the need for __________.
3. Notify the appropriate __________ __________ __________ through dispatch or direct contact.
4. Once on scene, meet with hazardous devices personnel to provide information about the __________ and __________ of the material or device.

Skill Drill 11-5: Identifying Samples and Evidence to Be Collected NFPA 1072: 6.5.1

Test your knowledge of this skill drill by filling in the correct words in the photo captions.

1. Note the __________ and __________ characteristics of any suspected hazardous materials evidence.
2. If possible, mark the locations with a(n) __________ __________ or color-coded identifier.

Skill Drill 11-6: Collecting Samples Using Equipment and Preventing Secondary Contamination NFPA 1072: 6.5.1

Test your knowledge of this skill drill by filling in the correct words in the photo captions.

1. The assistant holds open the evidence package or container so the sampler can place the evidence inside without __________-__________ the evidence.
2. The sampler obtains two samples: one sample to use for __________ __________ and another to preserve as evidence.

Skill Drill 11-7: Documentation of Sampling NFPA 1072: 6.5.1

Test your knowledge of this skill drill by filling in the correct words in the photo captions.

1. The documenter photographs and/or __________ the sampling and collection process.
2. The documenter notes the name of the person collecting the sample, the physical location of the agent, the state of the agent, the quantity present, the __________ of the sample collection, and the size and condition of the container.

Skill Drill 11-8: Labeling, Packaging, and Decontamination NFPA 1072: 6.5.1

Test your knowledge of this skill drill by filling in the correct words in the photo captions.

1. Seal the initial container with appropriate sealing tape, and place your initials on the tape or seal. This step will prevent __________.
2. Place the initial container in a secondary container, and seal this container in the same manner as the first container. Label the secondary container with a unique exhibit number, the name of the person who collected the item, and the location and time the sample was seized. Then, place the secondary container in a(n) __________ __________ so that the exhibit can pass through the decontamination procedure without being affected.

Operations Mission-Specific: Product Control

Workbook Activities

The following activities have been designed to help you. Your instructor may require you to complete some or all of these activities as a regular part of your hazardous materials training program. You are encouraged to complete any activity that your instructor does not assign you as a way to enhance your learning in the classroom.

Chapter Review

The following exercises provide an opportunity to refresh your knowledge of this chapter.

Matching

Match each of the terms in the left column to the appropriate definition in the right column.

	Term	Definition
_____	**1.** Vapor dispersion	**A.** The process of keeping a hazardous material within the immediate area of the release
_____	**2.** Confinement	**B.** Actions taken to stop a hazardous materials container from leaking or escaping its container
_____	**3.** Containment	**C.** The process of lowering the concentration of vapors by spreading them out
_____	**4.** Diversion	**D.** The process of controlling vapors by covering the product with foam or by reducing the temperature of the material
_____	**5.** Dilution	**E.** The process of adding some substance to a product to weaken its concentration
_____	**6.** Vapor suppression	**F.** Redirecting spilled material to an area where it will have less impact
_____	**7.** Retention	**G.** The process of applying a material that will soak up and hold the hazardous material
_____	**8.** Diking	**H.** Used when a liquid is flowing in a natural channel or depression, and its progress can be stopped by blocking the channel
_____	**9.** Damming	**I.** The placement of impervious materials to form a barrier that will keep a hazardous material in liquid form from entering an area
_____	**10.** Absorption	**J.** The process of creating an area to hold hazardous materials

CHAPTER

12

Multiple Choice

Read each item carefully, and then select the best response.

_____ 1. The process whereby a spongy material is used to soak up a liquid hazardous material is known as:
- A. absorption
- B. adsorption
- C. diversion
- D. retention

_____ 2. The process of redirecting the flow of a liquid away from an endangered area to an area where it will have less impact is known as:
- A. retention
- B. confinement
- C. diversion
- D. diking

_____ 3. The phase of a hazardous materials incident after the imminent danger to people, property, and the environment has passed or is controlled is referred to as:
- A. mitigation
- B. recovery
- C. termination
- D. demobilization

_____ 4. Which foam is used when large volumes of foam are required for spills or fires in warehouses, tank farms, and hazardous waste facilities?
- A. high-expansion
- B. protein
- C. fluoroprotein
- D. aqueous film-forming

_____ 5. A(n) __________ dam is placed across a small stream or ditch to completely stop the flow of materials through the channel.
- A. overflow
- B. underflow
- C. coffer
- D. complete

_____ 6. The process of attempting to keep the hazardous material on the site or within the immediate area of the release is known as:
- A. confinement
- B. containment
- C. exposure
- D. suppression

_____ 7. Most flammable and combustible liquid fires can be extinguished by using:
- A. water
- B. carbon monoxide
- C. foam
- D. dilution

_____ **8.** An MC-306/DOT-406 cargo tank is most likely to contain which one of the following materials?
A. corrosive
B. liquefied gas
C. flammable liquid
D. cryogen

_____ **9.** Why is the technique of absorption difficult for operational personnel to implement?
A. It creates an extensive clean-up process.
B. Absorbent materials are often difficult to obtain
C. It involves a large number of personnel.
D. It generally involves being in close proximity to the spill.

_____ **10.** The process of creating an area to hold hazardous materials is called:
A. retention
B. diking
C. damming
D. diversion

_____ **11.** The addition of another liquid to weaken the concentration of a hazardous material is called:
A. dispersion
B. dilution
C. extension
D. liquidation

_____ **12.** The process of lowering the concentration of vapors by spreading them out is called:
A. vapor suppression
B. vapor release
C. vapor evacuation
D. vapor dispersion

_____ **13.** When the imminent danger has passed and clean-up begins, the incident has reached the _______ phase.
A. demobilization
B. clean-up
C. recovery
D. termination

_____ **14.** Which phase of the incident includes the compilation of all records necessary for documentation of the incident?
A. administration
B. recovery
C. termination
D. demobilization

_____ **15.** Who typically makes the decision to end the emergency phase of a hazardous materials incident?
A. Safety Officer
B. Incident Commander
C. Operations Officer
D. Planning Officer

Vocabulary

Define the following terms using the space provided.

1. Alcohol-resistant concentrate

____

____

____

2. Underflow dam

____

____

____

3. Recovery phase

____

____

____

Fill-In

Read each item carefully, and then complete the statement by filling in the missing word(s).

1. The process of lowering the concentration of vapors by spreading them out is ____________________ ____________________.
2. A fog stream is commonly used to ____________________ vapors.
3. Highly volatile flammable liquids may be left to ____________________ on their own without taking offensive action to clean them up.
4. A protective action that should always be considered, especially with transportation or fixed facility incidents, is ____________________.
5. Fixed ammonia systems provide a good example of the effectiveness of using ____________________ ____________________.
6. ____________________ is the process of creating an area to hold hazardous materials.
7. A(n) ____________________ ____________________ is placed across a small stream or ditch to completely stop the flow of materials through the channel.
8. Alcohol-resistant concentrates are formulated so that polar solvents will not ____________________ the foam.
9. Most flammable and combustible ____________________ ____________________ can be extinguished by using Class B foam.
10. ____________________ refers to actions that stop the hazardous material from leaking or escaping its container.

True/False

If you believe the statement to be more true than false, write the letter "T" in the space provided. If you believe the statement to be more false than true, write the letter "F."

1. ______ Plugging a breached container is an example of containment.
2. ______ The opposite of absorption is adsorption.
3. ______ An underflow dam is used to contain materials with a specific gravity greater than 1.
4. ______ Firefighting foams should be sprayed directly on the burning material and surface.
5. ______ Sometimes, no action is the safest course of action.
6. ______ A public safety agency will often hand off the clean-up operations to a commercial clean-up company.
7. ______ MC 307/DOT 407 cargo tanks are certified to carry chemicals that are transported at high pressure.

8. _______ Many chemical processes, or piped systems that carry chemicals, have a way to remotely shut down a system or isolate a valve.

9. _______ Dilution can be used only when the identity and properties of the hazardous material are known with certainty.

10. _______ A retention technique is used to redirect the flow of a liquid away from an area.

Short Answer

Complete this section with a short written answer using the space provided.

1. List three types of firefighting foams.

___

___

___

___

Hazardous Material Alarms

The following real-case scenarios will give you an opportunity to explore the concerns associated with hazardous materials. Read each scenario, and then answer each question in detail.

1. You have been dispatched to a diesel fuel spill on the state route south of town. On arrival at the site, you find 10 gallons of diesel fuel spilled across the roadway. Your company officer directs you to absorb the fuel with on-board absorbent. How will you absorb the spilled hazardous material?

___

___

___

___

___

___

2. Before you can begin absorbing the spilled diesel fuel, the combustible material is ignited by a road flare placed too close to the spill.

a. What can you use to extinguish the burning combustible fuel?

b. How should the extinguishing agent be applied?

Skill Drills

Skill Drill 12-1: Using Absorption/Adsorption to Manage a Hazardous Materials Incident NFPA 1072: 6.6.1

Test your knowledge of this skill by filling in the correct words in the photo captions.

1. Decide which material is best suited for use with the spilled product. Assess the location of the spill and stay clear of any spilled product. Use detection and monitoring _______________, as well as reference sources, to identify the product. Apply the appropriate material to control the spilled product.
2. Apply the _______________ material to the spilled product.
3. Maintain _______________ of the absorbent/adsorbent materials and take appropriate steps for their disposal.

Skill Drill 12-2: Constructing an Overflow Dam NFPA 1072: 6.6.1

Test your knowledge of this skill by filling in the correct words in the photo captions.

1. Determine the need for, and location of, an overflow dam. Build a dam with _______________ or other available materials.
2. Install the appropriate number of _______________ to _______________ _______________ plastic pipes horizontally on top of the dam, and then add more sandbags on top of the dam. Complete the dam installation, and ensure that the piping allows the proper flow of water without allowing the heavier-than-water material to pass through the pipes.

Skill Drill 12-3: Constructing an Underflow Dam NFPA 1072: 6.6.1

Test your knowledge of this skill by filling in the correct words in the photo captions.

1. Determine the need for, and location of, an underflow dam. Build a dam with sandbags or other _______________ materials.
2. Install two to three lengths of 3- to 4-inch plastic pipes at a(n) _______________- to _______________-degree angle on top of the dam, and add more sandbags on top of the dam. Complete the dam installation, and ensure that the size will allow the proper flow of water underneath the lighter-than-water liquid.

Skill Drill 12-4: Constructing a Dike NFPA 1072: 6.6.1

Test your knowledge of this skill by filling in the correct words in the photo captions.

1. Determine the best location for the dike. If necessary, dig a depression in the ground 6″ to 8″ (15 cm to 20 cm) deep. Ensure that plastic will not _______________ adversely with the spilled chemical. Use plastic to line the bottom of the depression, and allow for sufficient plastic to cover the dike wall.
2. _______________ a short wall with sandbags or other available materials.
3. Complete the dike installation, and ensure that its _______________ will contain the spilled product.

Skill Drill 12-5: Using Dilution to Manage a Hazardous Materials Incident NFPA 1072: 6.6.1

Test your knowledge of this skill by filling in the correct word(s) in the photo caption.

1. Determine the _______________ of a dilution operation. Obtain guidance from a hazardous materials technician, specialist, or professional. Ensure that the water used will not overflow and affect other product-control activities. Add small amounts of water from a distance to dilute the product. Contact the hazardous materials technician, specialist, or other qualified professional if additional issues arise.

Skill Drill 12-6: Constructing a Diversion NFPA 1072: 6.6.1

Test your knowledge of this skill by filling in the correct words in the photo caption.

1. Determine the best location for the diversion. Use sandbags or other materials to divert the product flow to an area with _______________ _______________. Stay clear of the product flow. Monitor the diversion channel to ensure the integrity of the system.

Skill Drill 12-8 Using Vapor Dispersion to Manage a Hazardous Materials Incident NFPA 1072: 6.6.1

Test your knowledge of this skill by filling in the correct words in the photo captions.

1. Determine the viability of a dispersion operation. Use the appropriate monitoring instrument to determine the boundaries of a safe work area. Ensure that ________________. ________________. in the area have been removed or controlled.

2. Apply ________________. from a distance to disperse vapors. Monitor the environment until the vapors have been adequately dispersed.

Skill Drill 12-10: Performing the Rain-Down Method of Applying Foam NFPA 1072: 6.6.1

Test your knowledge of this skill drill by placing the following photos in the correct order. Number the first step with a "1," the second step with a "2," and so on.

______ Allow the foam to flow across the top of the pool of the product until it is completely covered.

______ Direct the stream of foam into the air so that the foam gently falls onto the pool of the product.

______ Open the nozzle to ensure that foam is being produced. Move within a safe range of the product and open the nozzle.

Operations Mission-Specific: Victim Rescue and Recovery

Workbook Activities

The following activities have been designed to help you. Your instructor may require you to complete some or all of these activities as a regular part of your hazardous materials training program. You are encouraged to complete any activity that your instructor does not assign you as a way to enhance your learning in the classroom.

Chapter Review

The following exercises provide an opportunity to refresh your knowledge of this chapter.

Matching

Match each of the terms in the left column to the appropriate definition in the right column.

	Term	Definition
______	**1.** Ambulatory victims	**A.** A crew of fully qualified and equipped responders who are assigned to enter the hot zone
______	**2.** Shelter-in-place	**B.** Individuals who function as a standby rescue crew of relief for those entering the hot zone
______	**3.** Entry team	**C.** Those who are able to walk on their own
______	**4.** Backup team	**D.** Sometimes this option is preferable to removing victims from a building
______	**5.** START	**E.** A triage system for large-scale, mass-casualty incidents

Multiple Choice

Read each item carefully, and then select the best response.

______ **1.** Which is the defining characteristic of recovery mode?
- **A.** The incident has been stabilized.
- **B.** Protective actions are fully implemented.
- **C.** There is no chance of rescuing a victim alive.
- **D.** All victims have been removed from the hot zone.

______ **2.** Which one of the following is NOT required when beginning a rescue in a hazardous materials situation?
- **A.** an entry team
- **B.** a backup team
- **C.** emergency decontamination staff
- **D.** a staging area

______ **3.** Which one of the following techniques CANNOT be used for a victim who is conscious and responsive but incapable of standing or walking?
A. two-person extremity carry
B. two-person seat carry
C. two-person chair carry
D. two-person walking assist

______ **4.** Which technique is particularly suitable when a victim must be carried through doorways, along narrow corridors, or up and down stairs?
A. two-person chair carry
B. two-person extremity carry
C. clothes drag
D. standing drag

______ **5.** When faced with a potential victim rescue, responders should first determine __________.
A. which exposures are at risk
B. if enough responders are on scene to make the attempt
C. the number of people involved
D. the type of material involved

______ **6.** What is the minimum number of team members that should be working in the hot zone?
A. 2
B. 3
C. 4
D. 5

______ **7.** In general, what is the minimum number of trained responders required to attempt a rescue in a life-threatening hazardous materials situation?
A. 4
B. 5
C. 6
D. 7

______ **8.** Which is the correct START triage category for a conscious victim of a hazardous materials exposure who is unable to follow commands?
A. Immediate
B. Delayed
C. Urgent
D. Minor

______ **9.** If an entry team consists of 4 personnel, what is the minimum size of the backup team?
A. 1
B. 2
C. 3
D. 4

______ **10.** Sorting victims of a mass casualty incident to establish priority for care is called __________.
A. evaluation
B. triage
C. assessment
D. prioritization

Labeling

Label the following diagram with the correct terms.

1. The START triage method.

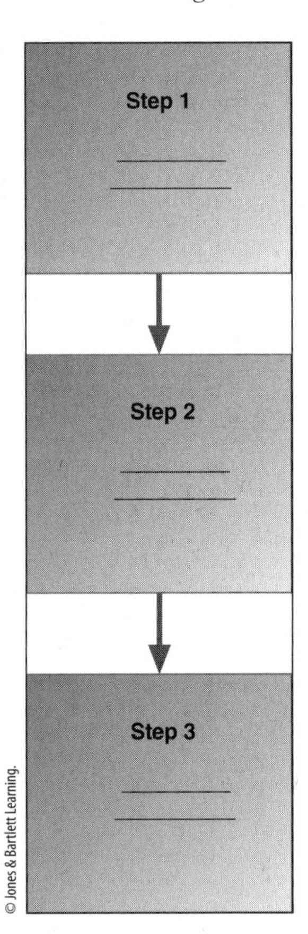

Vocabulary

Define the following terms using the space provided.

1. Rescue mode

____

____

____

2. Backup team

____

____

____

3. Triage

____

____

____

Fill-In

Read each item carefully, and then complete the statement by filling in the missing word(s).

1. ____________ victims are unable to walk under their own power.
2. When victims are present and determined to have a good chance for survival, the incident is considered to be in ____________ mode.
3. If there are six responders who will make up the entry team, there should be ____________ responders on the backup team.
4. The ____________-____________ ____________ ____________, also known as the sit pick, requires no equipment and can be performed in tight or narrow spaces.
5. An efficient method to rapidly move a victim from a dangerous location is to use a(n) ____________.
6. The ____________ is particularly suitable when a victim must be carried through doorways, along narrow corridors, or up or down stairs.
7. Whenever possible, a(n) ____________ ____________ should be used to remove a trapped victim from a vehicle.
8. OSHA requires that the minimum number of responders working in a hot zone be ____________.
9. Nonambulatory victims in mass-casualty situations must be sorted out based on their medical priority using an approved ____________ system.
10. When ____________ precedes a rescue, the entire process is more time-consuming and perhaps more complicated.

True/False

If you believe the statement to be more true than false, write the letter "T" in the space provided. If you believe the statement to be more false than true, write the letter "F."

1. ______ In most hazardous materials incidents, at least five trained responders are required to make a rescue attempt.
2. ______ In hazardous materials response, the entry team/backup team personnel ratio is always 1:1.
3. ______ Backup team members are dressed one PPE level down from that worn by the entry team.
4. ______ Typically, definitive medical care is rendered to victims during rescue mode.
5. ______ Emergency decontamination is performed in potentially life-threatening situations to rapidly remove the bulk of the contamination from an individual.
6. ______ Ambulatory victims who are within the line of sight and are able to walk may be directed and encouraged to leave the area under their own power.
7. ______ The cradle-in-arms carry can be used by one responder to carry a child or small adult.
8. ______ Whenever possible, a long backboard should be used to remove a victim trapped in a vehicle.
9. ______ Hazardous materials teams practiced the concept of rapid intervention before it was adopted in structural firefighting.
10. ______ Chemical gloves improve dexterity and make both clothing and skin easier to grasp.

Short Answer

Complete this section with short written answers using the space provided.

1. List the two emergency drags that can be used to remove unconscious victims from a dangerous situation.

___

___

___

___

2. Identify the acronym used in START triage, and describe the process.

___

___

___

___

Hazardous Material Alarms

The following real-case scenarios will give you an opportunity to explore the concerns associated with hazardous materials. Read each scenario, and then answer each question in detail.

1. You are the officer in charge of a fire company responding to a report of a chemical leak at a manufacturing plant adjacent to an expressway. The alarm was called in by a contract security guard who detected a sharp, pungent odor and found an outdoor 10,000-gallon chemical storage tank leaking into a secondary containment. When you arrive at the premises, you note there is no placard on the container; however, the ID number is clearly stenciled and visible from a distance. Because the material is spilling into a secondary containment, there is no immediate threat of the materials entering storm drains or the ground. The security guard called in from a cell phone but does not answer when your dispatcher attempts to contact him again for details. In his initial report, however, he stated that the material involved was methyl isocyanate. You can see that the security vehicle is parked alongside the tank and has its engine running and its yellow overhead lights flashing, and it appears that someone is lying on the ground next to the car. It is approaching 1:00 AM and a light rain has begun to fall. How can you safely rescue the downed security guard?

___

___

___

___

___

___

2. Your first ladder company reports that a second security guard has been found unconscious at a guard booth at the perimeter of the property. She is not breathing. What first-aid actions should the ladder company crew take while they await an ambulance?

___

___

___

___

___

___

Skill Drills

Skill Drill 13-1: Performing a Two-Person Walking Assist NFPA 1072: 6.8.1

Test your knowledge of this skill drill by placing the following photos in the correct order. Number the first step with a "1," the second step with a "2," and so on.

______ The responders assist the victim to a standing position.

______ Two responders stand facing the victim, one on each side of the victim.

______ Once the victim is fully upright, drape the victim's arms around the necks and over the shoulders of the responders, each of whom holds one of the victim's wrists.

______ Responders assist walking at the victim's speed.

______ Both responders put their free arm around the victim's waist, grasping each other's wrists for support and locking their arms together behind the victim.

Skill Drill 13-2: Performing a Two-Person Extremity Carry NFPA 1072: 6.8.1

Test your knowledge of this skill drill by placing the following photos in the correct order. Number the first step with a "1", the second step with a "2", and so on.

______ The second responder backs in between the victim's legs, reaches around, and grasps the victim behind the knees. arms.

______ Two responders help the victim to sit up.

______ The first responder kneels behind the victim, reaches under the victim's arms, and grasps the victim's wrists.

______ The first responder gives the command to stand and carry the victim away, walking straight ahead. Both responders must coordinate their movements.

Skill Drill 13-3: Performing a Two-Person Seat Carry NFPA 1072: 6.8.1

Test your knowledge of this skill drill by placing the following photos in the correct order. Number the first step with a "1," the second step with a "2," and so on.

_______ If possible, the victim puts his or her arms around the necks of the responders for additional support.

_______ Raise the victim to a sitting position, and link arms behind the victim's back.

_______ Kneel beside the victim near the victim's hips.

_______ Place your free arms under the victim's knees, and link arms.

Operations Mission-Specific: Response to Illicit Laboratories

Workbook Activities

The following activities have been designed to help you. Your instructor may require you to complete some or all of these activities as a regular part of your hazardous materials training program. You are encouraged to complete any activity that your instructor does not assign you as a way to enhance your learning in the classroom.

Chapter Review

The following exercises provide an opportunity to refresh your knowledge of this chapter.

Matching

Match each of the terms in the left column to the appropriate definition in the right column.

______	**1.** Methamphetamine	**A.**	Bacterial agent
______	**2.** Anthrax	**B.**	Viral agent
______	**3.** Smallpox	**C.**	Ricin
______	**4.** Toxin	**D.**	Used to manufacture, process, culture, or synthesize an illegal drug
______	**5.** Illicit laboratory	**E.**	Also known as crank or ice
______	**6.** Sulfur mustard	**F.**	Blood agent
______	**7.** Cyanide	**G.**	Choking agent
______	**8.** Chlorine	**H.**	Blister agent
______	**9.** LEL	**I.**	Explosive material specialists
______	**10.** Hazardous device personnel	**J.**	Determined by atmospheric monitoring

CHAPTER 14

Multiple Choice

Read each item carefully, and then select the best response.

______ **1.** Which is a potential chemical warfare agent?
A. anthrax
B. cyanide
C. Ebola
D. ricin

______ **2.** Which is a potential biological warfare agent?
A. sulfur mustard
B. anhydrous ammonia
C. botulinum
D. chlorine

______ **3.** Which is a common symptom of biological agent exposure?
A. difficulty breathing
B. nerve damage
C. dizziness
D. fever

______ **4.** Which is a common symptom of chemical agent exposure?
A. peripheral edema
B. hair loss
C. difficulty breathing
D. fever

______ **5.** Which of the following is a by-product of a living organism?
A. bacteria
B. fungus
C. virus
D. toxin

______ **6.** Lithium and sodium metal hazards are of particular concern at which type of laboratory?
A. Methamphetamine
B. Homemade explosives
C. Chemical warfare agent
D. Biological warfare agent

______ **7.** Which item is common to the various methamphetamine production processes?
A. protein supplements
B. sodium triphosphate (TSP)
C. cold medicine
D. castor beans

_____ **8.** Upon entering an illicit lab, you notice what appears to be an improvised explosive device. What should you do?
A. Exit the area immediately.
B. Saturate the device with water.
C. Mark and avoid the object while continuing operations.
D. Disconnect the battery leads.

_____ **9.** Which is a common source of red phosphorus used in illicit labs?
A. chemical drain uncloggers
B. matches
C. batteries
D. pool cleaners

_____ **10.** Ideally, direction in regard to an illicit lab incident should be provided by which of the following?
A. the fire department
B. unified command
C. the law enforcement agency having jurisdiction
D. a federal incident management team

Vocabulary

Define the following terms using the space provided.

1. Clandestine drug laboratory

____

____

____

2. Hazardous device personnel (bomb squad):

____

____

____

3. Illicit laboratory

____

____

____

Fill-In

Read each item carefully, and then complete the statement by filling in the missing word(s).

1. Red phosphorus labs and anhydrous ammonia labs are processes used to produce ____________________.

2. Detonating cords may look like ____________________ to an uninformed responder.

3. Symptoms of exposure to biological agents may take as long as ____________________ to become apparent.

4. Ultimately, an illicit lab is treated as a(n) ____________________ by investigators.

5. It is essential that whenever an explosive device is suspected to be involved, the appropriate ____________________ ____________________ personnel are notified.

True/False

If you believe the statement to be more true than false, write the letter "T" in the space provided. If you believe the statement to be more false than true, write the letter "F."

1. _______ A common characteristic of methamphetamine labs is the use of ephedrine and pseudoephedrine (cold medicine) tablets.
2. _______ Illicit laboratories can be small enough to fit inside the trunk of a car.
3. _______ Biological agents such as anthrax can be cultured illegally in illicit laboratories.
4. _______ Tablets are ground in household blenders as the first step in methamphetamine production.
5. _______ Clandestine drug laboratories will rarely have unusual chemical odors.
6. _______ Biological toxins are typically extracted from a plant or animal.
7. _______ Cyanide is typically extracted from the camphor bean, which gives it its characteristic smell.
8. _______ Evidence collection should begin after completion of site decontamination.
9. _______ Canine teams used in law enforcement may need to be decontaminated.
10. _______ Decontamination areas and equipment should be established prior to any responder entering an illicit laboratory.

Short Answer

Complete this section with a short written answer using the space provided.

1. List five methamphetamine chemicals and their legitimate uses.

____

____

____

____

Hazardous Material Alarms

The following real-case scenarios will give you an opportunity to explore the concerns associated with hazardous materials. Read each scenario, and then answer each question in detail.

1. You are the officer in charge of a fire company responding to a structure fire in an apartment building. On your arrival inside the premises, you note there is paraphernalia such as terrorist training manuals, propaganda, and documents associated with a known terrorist organization. Your crew also notes the presence of timers, switches, fuses, gunpowder, and what appears to be rescue rope with wires protruding from the end. What is your next course of action?

____

____

____

____

____

____

2. You need to assist in decontaminating law enforcement S.W.A.T. personnel. What should you advise and do for them?

____

____

____

____

____

____

Skill Drills

Skill Drill 14-1: Identifying Safety Hazards NFPA 1072: 6.9.1

Test your knowledge of this skill drill by filling in the correct words in the following statements.

1. Visually assess the structure or property that is suspected to contain a laboratory operation for outward warning signs, such as ____________________ - ____________________ ____________________, the presence of security or surveillance systems (including triggering devices and booby traps) unusual or extreme for the occupancy, precursor chemical containers, laboratory equipment, or hostile dogs or occupants. Establish a safe containment perimeter based on the hazards identified.
2. Notify the appropriate law enforcement personnel, technicians, and allied professionals based on the hazards identified. Make a(n) ____________________ of any victims who may be present and any symptoms they are reporting.

Skill Drill 14-2: Conducting Joint Hazardous Materials/Hazardous Device Team Operations NFPA 1072: 6.9.1

Test your knowledge of this skill drill by filling in the correct words in the following statements.

1. Discuss with appropriate personnel those materials or devices that are potentially ____________________ and/or ____________________.
2. Develop a joint response plan, if necessary, to render the device or materials safe for collection as evidence. This should include ____________________ ____________________ ____________________; perimeter support; detection and monitoring strategies, if applicable; a communication plan; emergency medical and evacuation plans; and the like.
3. Develop a(n) ____________________ plan to support all personnel and equipment.

Skill Drill 14-3: Decontaminating Tactical Law Enforcement Personnel NFPA 1072: 6.9.1

Test your knowledge of this skill drill by filling in the correct words in the following statements.

1. Provide clear instructions to the law enforcement officers entering the decontamination line. Realize that, although they may be ____________________ trained to wear their PPE, they may not be trained in decontamination procedures.
2. Instruct law enforcement officers to make any weapons safe by pointing them in a safe direction, ____________________ them completely, locking back the firing mechanism, and engaging the safety selector switch.
3. Consult with canine officers to determine the best way to handle the animal if decontamination is required. Instruct law enforcement officers handling a canine to maintain control of the animal during the entire process. Have them apply a(n) ____________________ to the animal if necessary.
4. Instruct law enforcement officers handling prisoners to maintain control of the prisoners at all times. It is recommended that ____________________ law enforcement officers control each prisoner during this process so that the officers can be decontaminated as well.
5. ____________________ decontamination procedures as necessary.

Operations Mission-Specific: Operating Detection, Monitoring, and Sampling Equipment

Workbook Activities

The following activities have been designed to help you. Your instructor may require you to complete some or all of these activities as a regular part of your hazardous materials training program. You are encouraged to complete any activity that your instructor does not assign you as a way to enhance your learning in the classroom.

Chapter Review

The following exercises provide an opportunity to refresh your knowledge of this chapter.

Matching

Match each of the terms in the left column to the appropriate definition in the right column.

	Term	Definition
______	**1.** CGI	**A.** A quick test carried out in the field to ensure the meter is operating correctly prior to entering a contaminated atmosphere
______	**2.** Bump test	**B.** Used to detect flammable and potentially explosive atmospheres
______	**3.** Photo-ionization detectors	**C.** Percentage of oxygen in ambient air at sea level
______	**4.** 20.9%	**D.** General survey instruments that detect vaporous chemicals at very low levels, even in the ppm range
______	**5.** 19.5%	**E.** Oxygen-enriched atmosphere
______	**6.** 23.5%	**F.** Oxygen-deficient atmosphere
______	**7.** Raman spectroscopy	**G.** Uses reagents to identify materials
______	**8.** Chemical test strips	**H.** Accounts for the different types of gases that might be encountered
______	**9.** Spectra	**I.** Uses a laser to identify materials
______	**10.** Relative response curve	**J.** The unique "fingerprint" of a material

Multiple Choice

Read each item carefully, and then select the best response.

______ **1.** From the time an air sample is drawn into the machine until the machine processes the sample and gives a reading is referred to as the:

- **A.** detection interval
- **B.** reaction time
- **C.** relative response factor
- **D.** calibration rate

CHAPTER 15

______ **2.** Which of the following is often referred to as "sewer gas"?
A. sulfur mustard
B. hydrogen sulfide
C. hydrogen
D. hydrogen cyanide

______ **3.** Which gas quickly deadens the sense of smell to the point where an individual may no longer be able to detect the gas by its odor?
A. sulfur mustard
B. hydrogen sulfide
C. hydrogen
D. hydrogen cyanide

______ **4.** Which device is appropriate for measuring flammable atmospheres at or below their LEL/LFL in air?
A. PID
B. GC
C. CGI
D. FID

______ **5.** The IDLH exposure limit for carbon monoxide is ______ ppm.
A. 35
B. 10,000
C. 150
D. 1200

______ **6.** In general, which type of hazard is the most dangerous?
A. inhalation
B. penetration
C. absorption
D. ingestion

______ **7.** Which is the process of setting or correcting a measuring device by adjusting it to match a known source?
A. calibration
B. initialization
C. zeroing
D. bumping

______ **8.** Which procedure is performed in the field prior to entering a contaminated atmosphere?
A. zeroing
B. calibration
C. setup
D. bump test

______ **9.** The ______________ is the time required for a detector to clear a previous reading so that a new reading can be taken.
A. reset time
B. clearing interval
C. recovery time
D. relative response rate

______ **10.** A device is __________ when it begins its operational period in a clean atmosphere by displaying normal values (or no values).

A. calibrated
B. zeroed
C. reset
D. cleared

Labeling

Label the following diagrams with the correct terms.

1. Types of detectors and monitors.

Courtesy of MSA – The Safety Company.

A. ______________________________

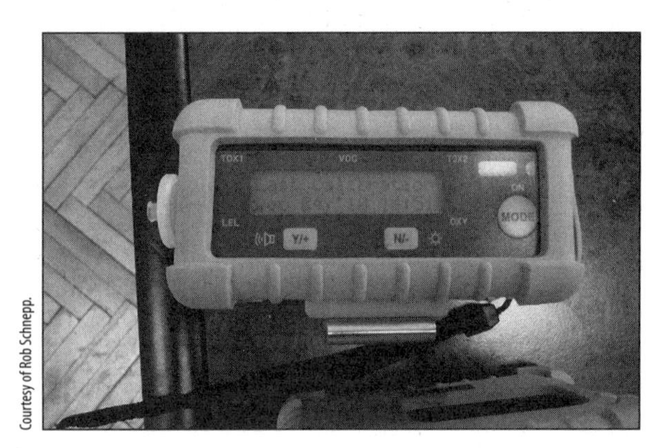

Courtesy of Rob Schnepp.

B. ______________________________

Courtesy of Rob Schnepp.

C. ______________________________

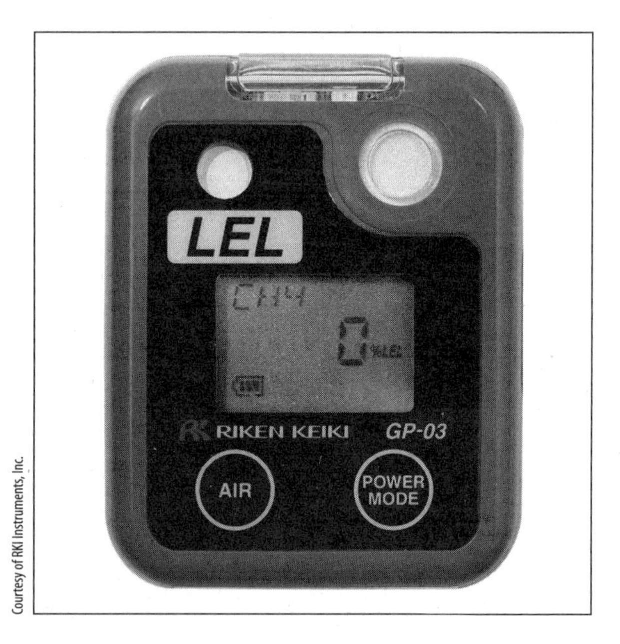

Courtesy of RKI Instruments, Inc.

D. ______________________________

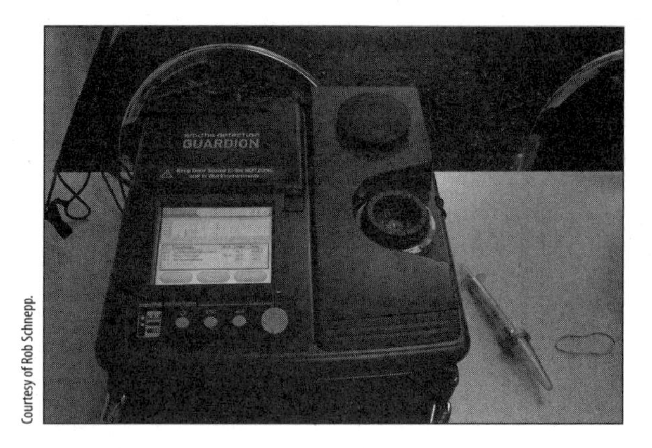

Courtesy of Rob Schnepp.

E. ______________________________

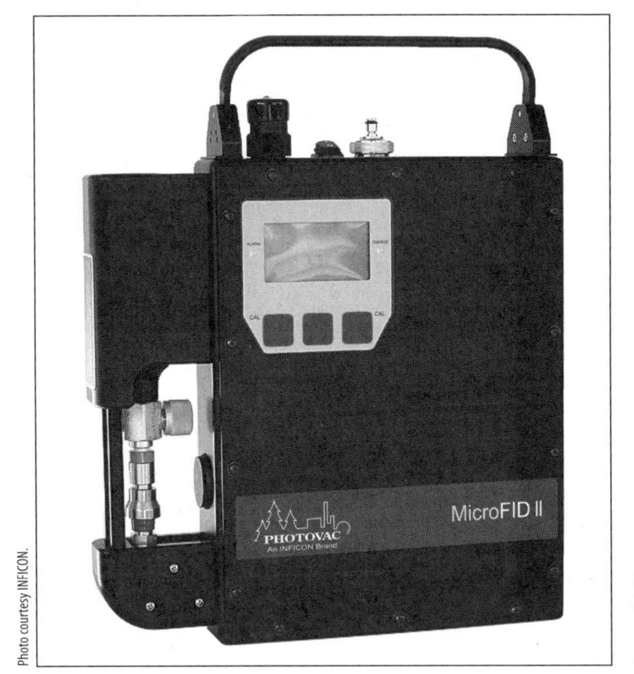

Photo courtesy INFICON.

F. ______________________________

Courtesy of Rob Schnepp.

H. 

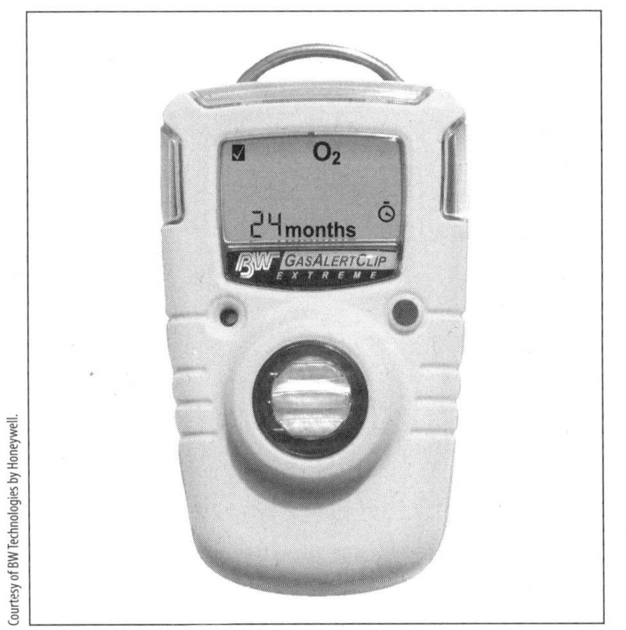

Courtesy of BW Technologies by Honeywell.

G. ______________________________

Courtesy of Sensit Technologies.

I. ______________________________

Vocabulary

Define the following terms using the space provided.

1. Volatile organic compound

____

____

____

2. Relative response curve

____

____

____

3. Situational awareness (SA)

____

____

____

Fill-In

Read each item carefully, and then complete the statement by filling in the missing word(s).

1. ____________________ is the process of ensuring that a particular instrument will respond appropriately to a predetermined concentration of gas.

2. __________ ______________ involves focusing on observing and understanding the visual clues available, orienting oneself to those inputs relative to the current situation, and making rapid decisions based on those inputs.

3. The manufacturer tests the monitor against various gases and vapors and provides a(n) __________ ________ ________ that can be used to determine the correct percentage of the gas being monitored.

4. A(n) _____ _____ _____ is a versatile instrument that can be used as either a general survey instrument or a qualitative instrument.

5. ______________ ____________________ is a chemical paper that allows the user to determine if a liquid or vapor is an acid or a base.

6. A device is "______________" when it begins its operational period in a clean atmosphere by displaying normal values (or no values) or when the device recovers to that same baseline state after exposure to a gas or vapor.

7. The ______________ ____________________ of a particular device is a function of how much time it takes a detector or monitor to clear so a new reading can be taken.

8. In atmospheres where the CO concentration is in the vicinity of ______________ ppm, all emergency personnel should be wearing SCBA.

9. ______________ ________________ ________________ look somewhat like pH paper but can perform several tests at once.

10. ______________-______________ ______________ are capable of detecting several different hazards at the same time.

True/False

If you believe the statement to be more true than false, write the letter "T" in the space provided. If you believe the statement to be more false than true, write the letter "F."

1. _______ The MSA CGI was the first and most popular monitoring instrument used in the fire service.
2. _______ Too little oxygen creates a health risk, but too much oxygen creates an elevated fire risk.
3. _______ The highest level of safety exists when the atmosphere reaches 100% of the LEL/LFL.
4. _______ Colorimetric tubes are designed to detect single substances and/or chemical families or groups.
5. _______ A carbon monoxide sensor may pick up the presence of hydrogen sulfide, hydrogen, or hydrogen cyanide.
6. _______ The pH scale goes from 0 (strong base) on the low end of the scale, to 14 (strong acid) on the opposing end of the scale.
7. _______ Electrochemical sensors have a shelf life of 5 to 7 years, depending on their use.
8. _______ In some cases, radiation detection devices can identify which type of radiation is present (alpha, beta, gamma).
9. _______ Personal dosimeters stay on the responder throughout an incident.
10. _______ Reaction time for a detection device should not exceed 5 to 10 seconds.

Short Answer

Complete this section with a short written answer using the space provided.

1. List the 10 basic actions for detection and monitoring.

____

____

____

____

Hazardous Material Alarms

The following real-case scenarios will give you an opportunity to explore the concerns associated with hazardous materials. Read each scenario, and then answer each question in detail.

1. You've responded to a spill in a chemistry laboratory at a local college. On arrival, you are alerted that there are five 1-gallon glass bottles that have broken and have spilled on the floor. The graduate student who reported the incident states that the material involved is tetrahydrofuran. He had attempted to clean it up himself for approximately 30 minutes, but stopped after experiencing irritated eyes, headache, nausea, and dizziness. The SDS indicates a flashpoint of 6°F (–14°C), an LEL of 2%, and a UEL of 11.8%. Your personnel indicate that they have measured 10% on their combustible gas indicator. In order to evacuate the building, security personnel pulled a manual fire alarm, which automatically shuts down the ventilation system as well. What potential explosion risk would you determine based on this scenario?

____

____

____

____

____

____

2. Describe the three steps of the start-up procedure for the CGI prior to entering the chemistry lab building.

____

____

____

____

____

____

Skill Drills

Skill Drill 15-2: Performing a Typical Start-Up Procedure for a Multi-Gas Meter NFPA 6.7.1

Test your knowledge of this skill drill by filling in the correct words in the photo captions.

1. Turn on the device and check the status of the _______. Allow the device to warm up (15 minutes is a reasonable amount of time in most cases, but you should always refer to the manufacturer's guidelines).
2. Ensure the device is "zeroed," or not picking up any readings while in a(n) __________ environment. Perform the appropriate bump test.
3. Allow the device to return to ________. The device is ready to use.

Skill Drill 15-3: Using a Multi-Gas Meter NFPA 1072: 6.7.1

Test your knowledge of this skill drill by filling in the correct words in the photo captions.

1. Turn the unit on and let it warm up (usually _____ minutes is sufficient). Ensure the battery has sufficient life for the operational period. Identify the installed sensors (e.g., oxygen, lower explosive limit, hydrogen sulfide, carbon monoxide) and verify they are not expired. (Note: Some units have a built-in PID function.) Review alarm limits and the audio and visual alarm notifications associated with those limits. Review and understand the types of gases and vapors that could harm or destroy the sensors. Use other methods to check for those substances (e.g., pH paper) to ensure they are not present in the atmosphere to be sampled. Care must be taken to avoid pulling liquids into the device—it is designed to sample air, not liquid!
2. Perform a test on the pump by ________________ the inlet and ensuring the appropriate alarm sounds.
3. Perform a fresh air calibration and "______" the unit.
4. Perform a bump test—expose the unit to a substance or substances that the unit should _______ and __________ to accordingly. In essence, you are making sure the unit will "see" what it is supposed to see before it might be called upon to see it in a real situation.
5. Allow the device to reset, or return to the __________ ________ _____ _______________, and then review the alarm levels and resetting procedures for addressing sensors that become saturated, or exposed to too much gas or vapor.
6. Review other device functions such as screen illumination, ________ _____________ (if available), and low battery alarm.
7. Review decontamination procedures. _____ _________________ monitoring and detection.

Skill Drill 15-4: Using Colorimetric Tubes NFPA 1072: 6.7.1

Test your knowledge of this skill drill by placing the following photos in the correct order. Number the first step with a "1," the second step with a "2," and so on.

______ Perform a pump test as specified by the manufacturer to ensure there are no leaks. A typical test is to use an unopened tube to hold the negative pressure inside the pump for a given period of time.

______ Collect all parts of the tube sampling system including pumps, tubes, extension hoses, etc.

______ Determine the mission and relevance of using colorimetric tubes for air sampling.

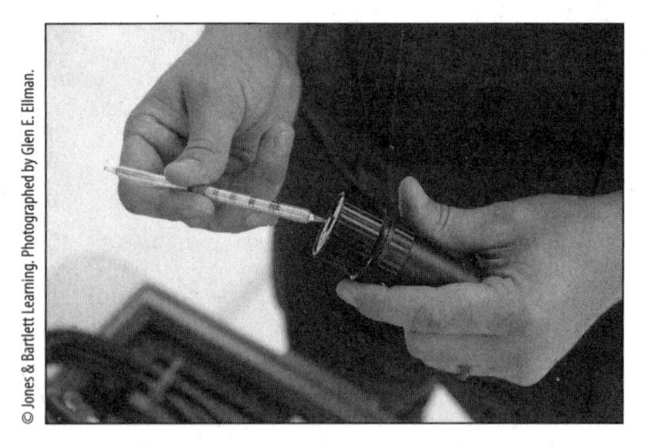

______ Assemble the tubes and other sampling apparatus prior to entering a contaminated area.

______ Select the appropriate tubes for the mission. Ensure the tubes are within their life expectancy—do not use outdated tubes.

______ Perform air sampling in accordance with the directions on the tube, using an unreacted tube as a control to view any color changes. Discard used tubes in accordance with manufacturer's guidelines.

______ Review the directions for using each tube, including pump stroke count, tube cross sensitivities, and expected reactions and color changes with the selected tubes.